ENERGY SCIENCE, ENGINEERING AND TECHNOLOGY SERIES

ENERGY COSTS, INTERNATIONAL DEVELOPMENTS AND NEW DIRECTIONS

ENERGY SCIENCE, ENGINEERING AND TECHNOLOGY SERIES

Oil Shale Developments
Ike S. Bussell (Editor)
2009. ISBN: 978-1-60741-475-9

Power Systems Applications of Graph Theory
Jizhong Zhu
2009. ISBN: 978-1-60741-364-6

Bioethanol: Production, Benefits and Economics
Jason B. Erbaum (Editor)
2009. ISBN: 978-1-60741-697-5

Introduction to Power Generation Technologies
Andreas Poullikkas
2009. ISBN: 978-1-60876-472-3

Handbook of Exergy, Hydrogen Energy and Hydropower Research
Gaston Pélissier and Arthur Calvet (Editors)
2009 ISBN: 978-1-60741-715-6

Energy Costs, International Developments and New Directions
Leszek Kowalczyk and Jakub Piotrowski (Editors)
2009 ISBN: 978-1-60741-700-2

ENERGY SCIENCE, ENGINEERING AND TECHNOLOGY SERIES

ENERGY COSTS, INTERNATIONAL DEVELOPMENTS AND NEW DIRECTIONS

LESZEK KOWALCZYK
AND
JAKUB PIOTROWSKI
EDITORS

Nova Science Publishers, Inc.
New York

For permission to use material from this book please contact us:
Telephone 631-231-7269; Fax 631-231-8175
Web Site: http://www.novapublishers.com

LIBRARY OF CONGRESS CATALOGING-IN-PUBLICATION DATA

Energy costs : international developments and new directions / [compiled by] Leszek Kowalczyk and Jakub Piotrowski.
p. cm.
Includes index.
ISBN 978-1-60741-700-2 (hardcover)
1. Renewable energy sources--Costs. 2. Power resources--Costs. I. Kowalczyk, Leszek. II. Piotrowski, Jakub.
TJ808.E584 2009
333.79--dc22

2009029199

Published by Nova Science Publishers, Inc. ✦ New York

CONTENTS

PREFACE

Energy is an essential factor in development, since it stimulates, and supports economic growth, and development. Any form of energy can be transformed into another form, but the total energy always remains the same. Several different kinds of energy exist to explain all known natural phenomena. These forms include (but are not limited to) kinetic, potential, thermal, gravitational, sound, light, elastic and electromagnetic energy. This book discusses new developments in this field, as well as the impact that renewable energy, and alternative energy might have in the future.

Chapter 1 - The environmental sustainability of the current global energy systems is under serious question. A major transition away from fossil fuels to one based on energy efficiency and renewable energy is required. Alternatively energy sources can potentially help fulfill the acute energy demand and sustain economic growth in many regions of Sudan. Renewable energy is beginning to gain importance in the global fight to prevent climate change. The development of a wide range of renewable energy technologies is continuing at an impressive rate. Cost reductions have been real and considerable, with few exceptions however, the relative contribution of renewable energy to the total energy mix remains quite small in Sudan. There are still a wide range of barriers preventing renewables from making a much more significant contribution. The increased exploitation of renewable energy sources is central to any move towards sustainable development. However, casting renewable energy thus carries with it an inherent commitment to other basic tenets of sustainability, openness, democraticisations, etc. This is an overview of some salient points and perspectives of renewable energy technologies in Sudan. This current literature is reviewed regarding the ecological, social, cultural and economic impacts of renewable technologies. This includes all the renewable energy technologies, energy conservation scenarios, energy savings and other mitigation measures necessary to reduce climate change. Sudan enjoys abundant solar, wind, hydro, and biomass energy resources. Sustainable low-carbon energy scenarios for the new century emphasise the untapped potential of renewable energy resources are required. Rural areas of Sudan can benefit from this transition. This chapter provides an overview of renewable energy activities and highlights future plans concerning optimum technical and economical utilisation of renewable energy available in Sudan.

Chapter 2 - The massive increases in fuel prices over the last years have however, made any scheme not requiring fuel appear to be more attractive and to be worth reinvestigation. In considering the atmosphere and the oceans as energy sources the four main contenders are wind power, wave power, tidal and power from ocean thermal gradients. The renewable energy resources are particularly suited for the provision of rural power supplies and a major advantage is that equipment such as flat plate solar driers, wind machines, etc., can be constructed using local resources and without the advantage results from the feasibility of local maintenance and the general encouragement such local manufacture gives to the build up of small-scale rural based industry. The key factors to reducing and controlling CO_2, which is the major contributor to global warming, are the use of alternative approaches to energy generation and the exploration of how these alternatives are used today and may be used in the future as green energy sources. Even with modest assumptions about the availability of land, comprehensive fuel-wood farming programmes offer significant energy, economic and environmental benefits. These benefits would be dispersed in rural areas where they are greatly needed and can serve as linkages for further rural economic development. Self-renewing resources such as wind, sun, plants and heat from the earth can provide clean abundant energy through the development of renewable technologies. Virtually all regions of the world have renewable resources of one type or another. Research and development investments in the past 25 years in renewable technologies development has lead to important advances in performance and resulting cost effectiveness. Renewable resources currently account for about 9%-10% of the energy consumed in the world; most of this is from hydropower and traditional biomass sources. Wind, solar, biomass and geothermal technologies are cost effective today in an increasing number of markets and are making important steps to broader commercialisation. The present situation is best characterised as one of very rapid growth for wind and solar technologies and of significant promise for biomass and geothermal technologies. Each of the renewable energy technologies is in a different stage of research, development and commercialisation and all have differences in current and future expected costs, current industrial base, resource availability and potential impact on energy supply. This article discusses the potential for such integrated systems in the stationary and portable power market in response to the critical need for a cleaner energy technology. Anticipated patterns of future energy use and consequent environmental impacts (acid precipitation, ozone depletion and the greenhouse effect or global warming) are comprehensively discussed in this paper. Throughout the theme several issues relating to renewable energies, environment and sustainable development are examined from both current and future perspectives.

Chapter 3 - Social learning is a prerequisite factor of technical innovation and diffusion, particularly in cases where personnel and financial support are limited like developing rural areas. This article sets out some considerations from social learning theory that offers potentials of stimulating renewable energy technology diffusion through active public participation in rural areas. Findings from the experiences of eco-village construction in China's Hainan province are used to illustrate how a governmental top-down initiative is reinforced by a "*farmer-to-farmer extension*" process in promoting the adoption of renewable energy technologies or equipments, such as anaerobic digesters and solar water heaters. By exploring who learns, what is learned, why it is learned, how it is learned and impacts of social learning occurring in the technology diffusion process of an eco-village in Hainan, this paper confirms that farmers' cognition of future reinforcements in environmental soundness and economic savings can lead to active social learning and adoption of renewable energy technologies. Financial support from the government can catalyze the growth of the awareness. Moving towards more positive social learning in technology diffusion in developing rural areas, emerging lessons from rural Hainan are combined with propositions concerning adaptive planning, public participation in community development projects, skills and relations of village leaders, technical specialists, accessibility to information and sources of technologies and stakeholder collaboration.

Chapter 4 - Growing concerns about social and environmental sustainability have led to increased interest in planning for the energy utility sector because of its large resource requirements and production of emissions. A number of conflicting trends combine to make the energy sector a major concern, even though a clear definition of how to measure progress toward sustainability is lacking. These trends include imminent competition in the electricity industry, global climate change, expected long-term growth in population and pressure to balance living standards (including per capital energy consumption). Designing and implementing a sustainable energy sector will be a key element of defining and creating a sustainable society. In the electricity industry, the question of strategic planning for sustainability seems to conflict with the shorter time horizons associated with market forces as deregulation replaces vertical integration. Sudan is an agricultural country with fertile land, plenty of water resources, livestock, forestry resources, and agricultural residues. Energy is one of the key factors for the development of national economies in Sudan. An overview of the energy situation in Sudan is introduced with reference to the end uses and regional distribution. Energy sources are divided into two main types; conventional energy (biomass, petroleum products, and electricity); and non-conventional energy (solar, wind, hydro, etc.). Sudan possesses a relatively high abundance of sunshine, solar radiation, and moderate wind speeds, hydro, and biomass energy resources. Application of new and renewable sources of energy available in Sudan is now a major issue in the future energy strategic planning for the alternative to the fossil conventional energy to provide part of the local energy demand. Sudan is an important case study in the context of renewable energy. It has a long history of meeting its energy needs through renewables. Sudan's renewables portfolio is broad and diverse, due in part to the country's wide range of climates and landscapes. Like many of the African leaders in renewable energy utilisation, Sudan has a well-defined commitment to continue research, development, and implementation of new technologies. Sustainable low-carbon energy scenarios for the new century emphasise the untapped potential of renewable resources. Rural areas of Sudan can benefit from this transition. The increased availability of reliable and efficient energy services stimulates new development alternatives. It is concluded that renewable environmentally friendly energy must be encouraged, promoted, implemented, and demonstrated by full-scale plant especially for use in remote rural areas.

Chapter 5 - Ground source heat pumps (GSHPs) provide an effective and clean way of heating buildings worldwide. They make use of renewable energy stored in the ground, providing one of the most energy-efficient ways of heating buildings. They are suitable for a wide variety of building types and are particularly appropriate for low environmental impact projects. They do not require hot rocks (geothermal energy) and can be installed in most of the world, using a borehole or shallow trenches or, less commonly, by extracting heat from a pond or lake. Heat collecting pipes in a closed loop, containing water (with a little antifreeze) are used to extract this stored energy, which can then be used to provide space heating and domestic hot water. In some applications, the pump can be reversed in summer to provide an element of cooling. The only energy used by the GSHP systems is electricity to power the pumps. Typically, a GSHP will deliver three or four times as much thermal energy (heat) as is used in electrical energy to drive the system. For a particularly environmental solution, green electricity can be purchased. The GSHP systems have been widely used in other parts of the world, including North America and Europe, for many years. Typically they cost more to install than conventional systems; however, they have very low maintenance costs and can be expected to provide reliable and environmentally friendly heating for in excess of 20 years. The GSHPs work best with heating systems, which are optimised to run at a lower water temperature than is commonly used in the UK boiler and radiator systems. As such, they make an ideal partner for underfloor heating systems. This article reviews seasonal thermal energy storage underground heat, and the potential of using heat pipe technology into the GSHP system. This research also covers soil thermal properties, estimated ground temperatures and methods of system control. It would appear to be a logical unit to rebate for a winter or summer peaking utility. The air heat pumps cannot match the lower cost of operating typical electric/gas equipment. The GSHP reduces demand and energy. Energy savings are sufficient enough to warrant added investments in the $600 to $900/ton range, especially if costs can be assumed in a 15 or 30-year mortgage.

Chapter 6 - The energy cost is one of the most important cost components in the water supply systems. Since large amounts of electricity are required to pump, transport and apply water, the profitability of some business, as irrigation districts or fishfarms, which use the water as production resource, is heavily dependent upon energy costs. Methodologies that can maximize energy cost savings while satisfying system performance criteria should be sought for the design and management of the water distribution systems. Some of these methods can be: (a) to improve the selection and/or operation of pumping stations; and (b) to include a regulating reservoir between the water supply source and the delivery system.

The alternative (a), related with the optimal design of pumping stations, refers to the selection of pump type, capacity and number of pump groups that results in minimum design and operating costs for a given water demand. With the alternative (b), if time-of-day energy tariff is available, the reservoir can be used for storing water that is pumped during off-peak hours to be used during peak hours. Also, the inclusion of reservoirs allows an equalization of pumping schemes that is not possible by pumping directly into the water delivery system. This allows pumping stations to operate near optimum energy efficiency.

In this chapter, one model that includes both alternatives was presented. To implement this optimization process, two algorithms were developed: (a) Algorithm for selection of least cost or optimum pump combinations in water supply systems and to evaluate the system's energy costs; and (b) Algorithm to determine the reservoir storage capacity that permits water to be pumped when energy tariffs are lowest and establish an annual pumping schedule in accordance with time-of-use energy tariffs. This model was applied to the water distribution systems of one irrigation district and one fishfarm located in southern Spain. The analysis indicated that the optimal selection of pump groups and/or the addition of a regulating reservoir was/were energy cost effective.

Chapter 7 - A brief review was given on the adsorption mechanism of methane. The effect of adsorption mechanism on separation and storage was interpreted. Methane as a supercritical gas adsorbs on solid surface merely monolayerly, and any novel material cannot assume the duty of methane/natural gas storage basing on adsorption. Instead, another mechanism was proposed for storage enhancement. Commentary was made on the research of methane separation with respects to the enrichment of methane or removal of carbon dioxide based on the adsorption mechanism.

Chapter 8 - During the past several years, world climate conditions have been abnormal. The G8 summit was held at Lake Touya in June 2008. At this meeting the G8 leaders agreed to make efforts to reduce CO_2 emissions to half of the present amount by 2050. The IPCC's studies of world energy consumption and emissions of CO_2 in 2050 are based on SRES-B CO_2, based on moderate world economic growth. CO_2 emissions are estimated to be 12.3Gt-C in 2050.

However, Asia's economic growth rate was higher than expected in recent years. At present, energy consumption in China in 2050 is expected to be 6Gtoe based on the past trend of Japan's and Korea's economic growth.

In the case of a 450ppm limitation, the allowable world CO_2 emission is about 4.8GtC by 2050. China's CO_2 emission is estimated to be 4.1Gt-C based on 6Gtoe energy consumption with the same IPCC t-C/toe ratio. A reduction of one-fifth of CO_2 emissions will be required in China by 2050.

There are two ways to reduce CO_2 emissions: one is reduction of energy consumption and another is the use of less CO_2 emission energy resources.

Concerning energy consumption, the ratio of energy consumption in Japan and China in the industrial field is about 1.5. Furthermore, it is required that energy efficiency will be improved two times the present rate of efficiency by 2050 in all fields of energy supply and consumption. The ratio of CO_2 emission per energy, t-C/toe, should be reduced to about half by substituting renewable energy (such as hydro, wind, photovoltaic, biomass), nuclear energy and CCS.

There are difficult barriers to achieving this target. The key components of this achievement are technologies in efficiency improvement and less CO_2 emissions in the energy supply at reasonable cost to assure the economic development. Also, the political stability and regimes are essential matters domestically and internationally. The climate crisis is a critical issue for mankind. It is expected that China will play a significant leadership role in the world.

Chapter 9 - Dye-sensitized solar cells (DSCs) considered as a promising alternative to conventional silicon-based photovoltaics have been intensively studied for the recent two decades. Although satisfactory solar energy conversion efficiencies based on such nanocrystalline mesoscopic devices have been achieved in laboratory, large-scale application is still in its infancy. DSCs mostly employ liquid electrolytes which facilitate charge transfer and ensure high energy conversion efficiency. However, evaporation and leakage of the organic solvents in liquid electrolytes drastically decrease the long-term stability and make out-door operation difficult to endure. Quasi-solid and solid-state DSCs have been designed and investigated to solve this problem. Nonetheless, there exists a significant disparity in photovoltaic performances of these devices compared to liquid electrolyte-based cells, mainly attributed to less facile charge transfer in quasi-solid and solid-state electrolytes.

In this short communication, a concept of oriented one-dimensional nano-channel structure enlightened by our previous results was proposed to facilitate charge transfer in DSCs. The application of such a concept in further promoting energy conversion efficiency of quasi-solid and solid-state DSCs is also discussed.

In: Energy Costs, International Developments...
Ed: Leszek Kowalczyk and Jakub Piotrowski

ISBN: 978-1-60741-700-2

Chapter 1

OPPORTUNITIES FOR RENEWABLE ENERGY IN RURAL COMMUNITIES OF SUDAN

Abdeen M. Omer, Siddig Adam and Khalid S. Ahmed
17 Juniper Court, Forest Road West, Nottingham NG7 4EU, UK

ABSTRACT

The environmental sustainability of the current global energy systems is under serious question. A major transition away from fossil fuels to one based on energy efficiency and renewable energy is required. Alternatively energy sources can potentially help fulfill the acute energy demand and sustain economic growth in many regions of Sudan. Renewable energy is beginning to gain importance in the global fight to prevent climate change. The development of a wide range of renewable energy technologies is continuing at an impressive rate. Cost reductions have been real and considerable, with few exceptions however, the relative contribution of renewable energy to the total energy mix remains quite small in Sudan. There are still a wide range of barriers preventing renewables from making a much more significant contribution. The increased exploitation of renewable energy sources is central to any move towards sustainable development. However, casting renewable energy thus carries with it an inherent commitment to other basic tenets of sustainability, openness, democraticisations, etc. This is an overview of some salient points and perspectives of renewable energy technologies in Sudan. This current literature is reviewed regarding the ecological, social, cultural and economic impacts of renewable technologies. This includes all the renewable energy technologies, energy conservation scenarios, energy savings and other mitigation measures necessary to reduce climate change. Sudan enjoys abundant solar, wind, hydro, and biomass energy resources. Sustainable low-carbon energy scenarios for the new century emphasise the untapped potential of renewable energy resources are required. Rural areas of Sudan can benefit from this transition. This chapter provides an overview of renewable energy activities and highlights future plans concerning optimum technical and economical utilisation of renewable energy available in Sudan.

Keywords: Sudan; renewable energy technologies; hydropower; biomass energy; geothermal power; greenhouses applications.

1. Introduction

Energy is an essential factor in development since it stimulates, and supports economic growth, and development. Fossil fuels, especially oil and natural gas, are finite in extent, and should be regarded as depleting assets, and efforts are oriented to search for new sources of energy. The clamour all over the world for the need to conserve energy and the environment has intensified as traditional energy resources continue to dwindle whilst the environment becomes increasingly degraded. Biomass energy supply in Sudan contributed 87% of total energy supply since 1980's. This includes firewood, charcoal and crop residues. Out of total fuel wood and charcoal supplies 92% was consumed in household sector with most of firewood consumption in the rural areas. Biogas, biofuels and woody biomass are other forms of energy sources that can be derived from organic waste materials.

In 1982, the Sudanese government had issued a decree, which made the National Electricity Corporation (NEC) as a statutory corporation under the Ministry of Energy and Mining (MEM). The NEC is responsible of the electricity generation, transmission and distribution in Sudan, and has the goal of improving electricity system network in order to achieve sustainability. Within this broad aim, this search has specific goals, including:

- To involve a wide-range of electricity sector participants, including utilities, regulators, environmentalists, and customers.
- To identify and implement a wider range of measures that will be direct indicators of sustainability than the current response impacts and pollutant outputs.
- To create improved electricity sector modelling tools that simulate electricity system operation under competition, and include transmission and distribution effects more directly into strategic planning.
- To consider technology options beyond the current generation.
- To increase the comprehensiveness of current life-cycle assessment, by including broader range of technologies, inputs and outputs are sensitive to technology choice versus the state of underlying production and transport infrastructures.
- To improve decision analysis tools to assist stakeholders in reaching consensus on complex alternatives, option portfolios, and flexible contingencies.
- To make the data and models developed for understanding the relationship between power systems and sustainability, available and usable through internet access.

Sudan's electricity sector is overwhelmed by poor infrastructure and frequent outages. This situation has witnessed dramatic change over the last three years. Many parts of Sudan now enjoying a steady flow of electricity, the development occurred concurrently with many managerial reforms, one of them was the launch of a journey to implement a total quality management (TQM) programme in 1996 [1].

NEC handled wide variety of electricity generation technologies (Table 1), including:

- Thermal (fossil, combined cycle, and combustion turbines).
- Storage (pumped hydro, batteries, and compressed air).
- Non-dispatchable technologies (solar, wind, cogeneration, and load management).

Table 1. Electricity output of present power plants in Sudan (GWh) [1]

Plants	Power (GWh)
Hydro power stations	1095.12
Steam power stations	1037.93
Diesel power stations	359.9
Gas turbine power stations	235.35
Combined power stations	1020.62
Total	3748.9
Peak load for national grid (MW)	534
Total number of consumers (head)	10700000
Total number of employees (head)	7097

The privatisation and price liberalisation in energy fields has been secured to some extent (but not fully). Availability and adequate energy supplies to the major productive sectors is still needed. The result is that, the present situation of energy supplies is for better than ten years ago. The investment law has also encouraged the participation of the investors from the national level as well as from the international friendly and sisters' countries to invest in energy supply systems such as:

- Petroleum products (import in particular) in the northern states.
- Electricity generation (in some states) using large diesel engine driven generators.

The readily implementation of electricity price liberalisation has some to extent relieved the National Electricity Corporation (NEC) from the heavy dependency on the government subsidies, and a noticeable improved of the NEC management, and electricity supplies are achieved.

Sudan aims to increase its electricity generation capacity over the next few years. In 2003, Sudan had 760 megawatts (MW) of electricity generation capacity, and total electricity generation of 3.2 billion KWh. The country's main generating facility is the 280-MW Roseires dam located on the Blue Nile river basin, approximately 315 miles southeast of Khartoum. Rebel groups have frequently attacked the facility, and the low water levels often cause its capacity to fall to 100 MW.

In recent years, Sudan has increased efforts to exploit renewable energy sources and reduce its dependence on oil. Wind, solar and biomass offers a variety of renewable options that are well suited to the African climate. A number of renewable energy initiatives are under way in Sudan that can contribute to rural development while also addressing climate change mitigation.

2. Hydropower

Sudan has an area of 2,505,813 km^2 and a population of about 38.5 million, more than 50 per cent of whom live in rural areas. The annual growth rate is 2.8 per cent and population density is 14 persons per square km in 2009. Sudan has a plenty of water resources to produce electricity. An overview of the electricity situation in Sudan is introduced with reference to hydropower potential and distribution. Hydropower utilisation in Sudan is promising. Sustainable low-carbon energy scenarios for the new century emphasise that the untapped potential of hydro-energy resources is required. The increased availability of reliable and efficient energy services stimulates new development alternatives. A number of prospective areas have been identified through surveys and studies carried for exploration of mini-hydropower resources in Sudan [1]. It is concluded that hydropower is environmentally friendly energy and must be encouraged, implemented and demonstrated by full-scale plant for rural electrification.

The electricity network is formed through two interconnected electrical grids, the Blue Nile grid and the Western grid, which cover only a small portion of the country [1]. Many regions of Sudan are not covered by the national grid, and rely on small diesel engine driven generators, although blackouts and brownouts are common (Figure 1). Only 30% of the population currently has access to electricity, although the government hopes to increase this figure by up to 90% in the coming years.

2.1. Water Resources

The annual average rainfall ranges from about 1 mm in the northern desert to about 1600 mm in the southern equatorial region. The total annual rainfall is estimated at 1093.2 x 10^9 m^3. The Ministry of Irrigation and Water Resources (MIWR) is in charge of water resources. Other authorities, which are responsible for water resources include the National Water Corporation (NWC). The 25 State Water Corporations–Ministry of Engineering Affairs of States, and four other private companies including the National Company for Manufacturing Water Equipment Limited (NCMWE), the National Drilling and Investment Company (NDIC), the National Water Resources Development (NWRD), and the Administration of Groundwater Research (AGR). Per capita domestic water consumption is estimated at 20 litres/day in urban areas and 5 litres/day in rural areas. The total water storage available at all dams in Sudan is about 18 km^3. There are four large dams in operation, three embankments and one concrete structure. The 60 m high Merowe concrete faced rockfill dam is another addition to hydropower in Sudan [2].

2.2. Hydropower Development

The Ministry of Energy and Mining (MEM) is in charge of the energy sector in general, and recently established Ministry of Electricity (ME) is now in charge of electricity generation. The National Electricity Corporation (NEC), which operates under the ME owns and operates hydro plants, isolated diesel systems, and thermal and steam plants. Primary energy consumption was 8500 x 10^6 MWh in 2002. Sources of primary energy were: non-wood fuel (9.2 per cent), wood fuel (77.8 per cent), petroleum (12 per cent) and electricity 1 per cent). Final energy consumption was 6500 x 10^6 MWh and was derived from wood fuel (77.8 per cent), electricity (7.3 per cent), petroleum (6.5 per cent) and non-wood fuel (8.4 per cent). Hydropower supplied 55 per cent of electricity. The remaining 45 per cent of electricity was produced from imported thermal energy. Total electricity consumption in 2002 was 640 x

10^3 MWh, representing per capita consumption of 53 kWh/year. It is estimated that total energy consumption will increase by 30 per cent over the next decade, and electricity demand by 70 per cent. Sudan's installed capacity is 640 MW, of which about 520 MW on the grid is available due to losses in transmission and problems at the dams. The country has a generation and transmission programme, according to which total capacity will increase to 3383 MW in 2016. In Sudan, electricity costs an average of 20 Sudanese Pound per kWh (US$ 1 = 264 Sudanese Pound). Four major hydro plants are in operation: Roseires (275 MW) and Sennar (15 MW), both on the Blue Nile, Rumela (20 MW) and Khashm El Girba (13 MW), on the Atbara River. The government has now decided to go ahead with the 1250 MW Merowe project on the River Nile, about 350 km north of Khartoum, and the scheme was put out to tender in 2001. The project will include a 60 m-high, 4800 m-long concrete faced rockfill dam (CFRD), and the powerhouse will contain ten 125 MW units (Table 2).

2.3. Growth of Electricity Sector

Several projects are underway to increase the Sudanese generating capacity. The largest projects include the proposed 1250-MW Merowe and 300-MW Kajbar hydroelectric facilities in the northern Sudan. France's Alstom, China's Harbin Power and several Arab investors have contributed in funding the construction of the Merowe facility, which was completed in July 2008. China is financing 75% of the $200 million Kajbar dam construction, and Sudan government providing the remaining 25%. Environmental groups have expressed concern about the Kajbar project, citing potential damage to the Nile ecosystem and the culture of displaced Nubian residents of the area.

In addition, to the Merowe and Kajbar facilities, Sudan inaugurated two electric power stations in June 2004. These stations are estimated to have a combined capacity of 330 MW. In November 2004, Sudan's first independent power production (IPP) project (257 MW) came online. The diesel plant, located near Khartoum, sells output to the state-owned Sudan Electricity Corporation (SEC). Several additional power stations with a total capacity of 700 MW are scheduled for completion before 2015.

Foreign investment in the Sudanese power sector is expected to increase with the cessation of the civil war. In June 2004, for example, the United Arab Emirates (UAE) pledged to invest in the Sudanese power sector following the signing of a peace agreement. In January 2006, the Export/Import Bank of India extended $350 million of credit to Sudan. The money was used for constructing a 500 MW power plant, which was built by India's Bharat Heavy Electricals Ltd.

2.4. Small Hydropower

Small hydro power potential in Sudan is also promising. A number of prospective areas have been identified by surveys, and studies, which are being carried out to explore mini hydro resources using waterfalls with heads ranging from 1 to 100 m. Also, the current flow of the River Nile water could be used to run in-stream turbines; water could then be pumped to riverside farms. There are more than 200 suitable sites for the use of in-stream turbines along the Blue Nile and the main River Nile. The total potential of mini hydro is estimated 67

GWh/year for the southern region, with 3785 MWh/year in the Jebel Marra, and 44895 MWh/year in El Gezira and El Managil canals. About 8500 kW of hydropower capacity is planned at the three plants.

Table 2. Feasible hydropower projects in Sudan

River	Station	Installed capacity (10^3 kW)	Energy (10^6 kWhy^{-1})
Main Nile	Dal	780	3150
	Merowe	1000	3965
	Mugrat	240	1360
	Dugash	285	1520
	Shereik	350	2000
	Sabaloka	120	520
White Nile	Jebel Aulia	35	150
Blue Nile	Rosaries	380	1590
	Sennar	45	150
Atbara	Khashm El Girba	12.5	40
	Rumela	30	100
Bahr El Jebel	Fola	720	3900
	Shekaila	210	1430
	Laki	210	1500
	Beddan	400	2600
	Juba	10	40
Tributaries	Sue	12	45
	Yei1	3	10
	Yei2, 3,4	0.3	2
	Kenati1-4	3.5	15
	Kaia	13.5	50
Total		4860	24137

A study is under way on a 400 km-long 500 kV transmission line from Merowe to Khartoum. There is about 520 MW of hydro capacity in operation, of a total power plant capacity of 640 MW. No mini hydro plants are under construction, but up to 2000 MW could be developed in the future. There are three plants with a capacity greater than 10 MW, and three plants are part of multi-purpose developments. There is a very large potential (about 3000 MW) for increasing the capacity at existing hydropower sites. Refurbishment is currently under way on the 40 MW units at Roseires plant.

A contract was recently awarded by NEC to VA TECH hydro to install 80 matrix generating sets, in 40 modules, at Jebel Aulia dam on the White Nile, about 40 km south of Khartoum. The total capacity will be 30.4 MW. Jebel Aulia dam was originally constructed for irrigation. Large-scale, conventional, power plant such as hydropower has an important part to play in development. It does not, however, provide a complete solution. There is an important complementary role for the greater use of small-scale, rural based-power plants.

2.5. Environment

All projects are required to have environmental impacts assessment conducted, covering all potential damage to the environment, mitigation and restoration, a reclamation plan including a resettlement programme for displaced residents, and the estimated implementation costs. All hydro projects are required to conduct an environmental impact study. The National Council for Environment has been established to develop a legal framework and policy of the Federal Environmental Protection since 1990. There is a legal framework in the country covering environmental impact studies for water resource development projects, and local people are informed about plans for new or refurbishment schemes through newspapers, radios, and television.

2.6. Future Outlook

In the past, Sudan was badly affected by rapidly rising fuel costs, but the country is now in a position to export oil. The country intends to implement new hydro projects to meet increasing demand and to avoid power shortages. Small-scale, decentralised hydropower can provide valuable reserve power and potentially make major contributions to local energy needs, especially the southern region along the White Nile from Malakal to the border with Uganda. There are plans to develop other sources of renewable energy in the next five to ten years, including solar energy: solar photovoltaic and solar thermal, wind energy, biomass, mini hydro and geothermal energy.

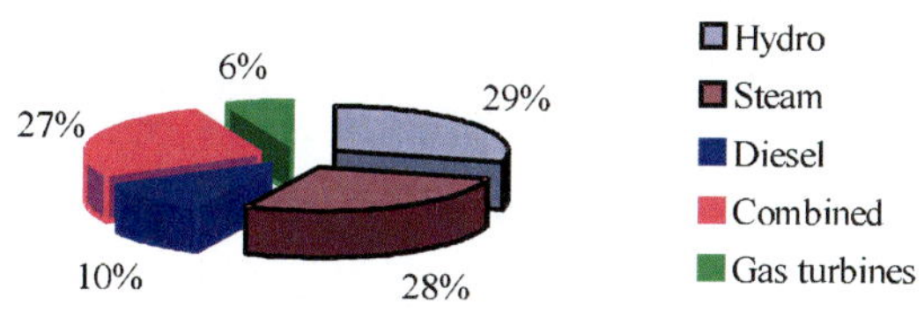

Figure 1. Electricity generated in 2005 (GWh)

In a country with a dense population, there are extreme pressures on energy supply system, which can stunt the country's economic growth. However, Sudan has recognised the potential to alleviate some of these problems by promoting hydro energy and utilising its vast and diverse climate, landscape and resources and by enhancing these for energy production. Thus, Sudan may stand at the forefront of the global hydro energy community and present an example in this field.

Energy efficiency brings health, productivity, safety, comfort and savings to the homeowner, as well as local and global environmental benefits. The use of hydro energy resources could play an important role in this context, especially with regard to responsible and sustainable development. It represents an excellent opportunity to offer a higher standard of living to the local people and will save local and regional resources. Implementation of hydro energy offers an opportunity for economic improvement by creating a market for producing companies as well as maintenance and repair services. It is expected that the pace of implementation will increase and the quality of work and also improve in addition to building the capacity of the private and district staff in contracting procedures.

3. Biomass Energy Potential

Agriculture is the backbone of economic and social development in Sudan. Biomass resources in Sudan play a significant role in energy supply as shown in Tables 3 and 4. Cooking is largely done using firewood (45%) and charcoal (30%) [1]. Hence, 75% of total energy per annum, representing roughly 3 million metric tons of forest reserves, and agricultural residues comes mainly from cotton stalks, groundnut shells and bagasse with estimates of more than 15 million metric tons. Also, considerable amount of non-woody biomass available is animal dung, estimated 17 million tons as shown in Table 5. Water hyacinth and aquatic weeds are estimated 9000 and 3000 tons per annum respectively [1]. Biomass residues are more economically exploitable and more environmentally benign than dedicated biomass resources. There exists a variety of readily available sources in Sudan, including agricultural residues such as sugar cane bagasse, and molasses, cotton stalks, groundnut shells, tree/forest residues, aquatic weeds, and various animal wastes [3].

Table 3. Sources of biomass energy available in Sudan 10^6 tons of equivalents (TOE)

Item	Source	10^6 TOE
1.	Natural, and cultivated forests	2.90
2.	Agricultural residues	6.20
3.	Animal wastes	1.05
4.	Water hyacinth	3.16
Total		13.31

Table 4. Biomass energy consumption in Sudan 10^3 tons of equivalents (TOE) [28]

Item	Sector	10^3 TOE	(%)
1.	Residential	4549	92.0
2.	Industrial	169	3.4
3.	Others*	209	4.6
Total		4927	100.0

*Others are commercial, constructions, and Quranic schools.

Table 5. Biomass energy potential from animal dung in different states of Sudan [28]

Item	States	Animal dung available (10^3 Tons)	Energy (TOE)
1.	Northern states	102.4	1543
2.	Eastern states	1222.9	18431
3.	Khartoum state	104.3	1572
4.	Central states	4223.7	63658
5.	Darfur states	5062.5	36301
6.	Kordofan states	2596.9	79140
7.	Southern states	4545.2	68505
Total		17857.9	269150

3.1. Bioenergy

Bioenergy is energy from the sun stored in materials of biological origin. This includes plant matter and animal waste, known as biomass. Plants store solar energy through photosynthesis in cellulose and lignin, whereas animals store energy as fats. When burned, these sugars break down and release energy exothermically, releasing carbon dioxide, heat and steam. The by-products of this reaction can be captured and manipulated to create power, commonly called bioenergy. Biomass is considered renewable because the carbon is taken out of the atmosphere and replenished more quickly than the millions of years required for fossil fuels to form. The use of biofuels to replace fossil fuels contributes to a reduction in the overall release of carbon dioxide into the atmosphere and hence helps to tackle global warming.

The range of waste treatment technologies that are tailored to produce bioenergy is also growing. There are a number of key areas of bioenergy from wastes including (but not limited to) biogas, biofuels and bioheat. When considering use of bioenergy, it is important to take into account the overall emission of carbon in the process of electricity production.

Scientific assumptions for a short-term energy strategy suggest, that the most economically efficient way to replace the thermal plants is to modernise existing power plants to increase their energy efficiency and to improve their environmental performance. Although a lack of generating capacity is forecasted in the long-term, utilisation of the existing renewable energy potential and the huge possibilities for increasing energy efficiency are sufficient to meet future energy demands in Sudan in the short-term (Table 6).

A total shift towards a sustainable energy system is a complex and long process, but is one that can be achieved within a period of about two decades. Implementation will require initial investment, long-term national strategies and action plans. However, the changes will have a number of benefits including:

- A more stable energy supply than at present.
- An improvement in the environmental performance of the energy sector.
- Certain social benefits.

A vision used a methodology and calculations based on computer modelling that utilised:

- Data from existing governmental programmes.
- Potential renewable energy sources and energy efficiency improvements.
- Assumptions for future economy growth.
- Information from studies and surveys on the recent situation in the energy sector.

In addition to realising the economic potential identified by the National Energy Savings Programme, a long-term effort leading to a 3% reduction in specific electricity demand per year after 2020 is proposed. This will require:

- Further improvements in building codes.
- Continued information on energy efficiency.

The environmental non-governmental organisations (NGOs) in Sudan are urging the government to adopt sustainable development of the energy sector by:

- Diversifying of primary energy sources to increase the contribution of renewable and local energy resources in the total energy balance.
- Implementing measures for energy efficiency increase at the demand side and in the energy generation sector.

Table 6. Estimate of gross domestic product by economic activity (10^6 Sudanese Dinar) [4]

Sectors	79-84	85-95	96-2000
Agricultural	18661 (43%)	77342 (40%)	230923 (41%)
Commercial	5888 (14%)	26522 (14%)	83472 (15%)
Manufacturing and mining	2652 (6%)	13514 (7%)	43069 (8%)
Transport and communication	4323 (10%)	19391 (10%)	57984 (10%)
Construction and public works	1917 (4%)	9879 (5%)	11200 (2%)
Electricity and water	664 (2%)	3916 (2%)	12340 (2%)
Government services	3322 (8%)	16853 (9%)	42095 (8%)
Personal services	1152 (3%)	5386 (3%)	17158 (3%)
Finance and real estate	4324 (10%)	20120 (10%)	60448 (11%)
Total	42903 (100%)	192923 (100%)	558688 (100%)

Table 7. Forestry resources in Sudan [5]

Regions	Total area (10^3 Hectares)	Average volume (m^3)	Total volume (m^3)
Northern	0	0	0
Eastern	2748	3	7008
Central	5124	5	25532
Khartoum	5	60	300
Kordofan	11628	11	122828
Darfur	17693	27	477200
Southern	71096	19	1332953
Total	108294	125	1965821

The price of natural gas is set by a number of market and regulatory factors that include:

- Supply and demand balance based on fuel market principles.
- Weather.
- Pipeline availability and deliverability.
- Storage inventory.
- New supply sources.
- Prices of other alternatives energy sources.
- Regulatory issues and uncertainty.

Recent development due to rehabilitation and improvement in the agricultural sector has raised the share to 41% as shown in Table 6. About 70% of this is located in the southern region, and the forestry area decrease towards the north of Sudan (Table 7). The available biomass resources including agricultural, aquatic weeds and various animal wastes are shown in Tables 8-10.

Table 8. Annual total different biomass supply in regions of Sudan (10^3 kg)

Regions	Agricultural residues	Animal waste	Bagasse
Northern	288	8548	8644
Eastern	11530	84316	84414
Central	21550	107342	131744
Khartoum	0	5766	5170
Kordofan	2733	234086	236819
Darfur	6901	173376	174067
Southern	1464	290599	197875
Total	44466	904033	838733

Table 9. Per capita consumption for household (10^3 kg)

Fuel	Wood		Charcoal	
Region	Urban	Rural	Urban	Rural
Northern	0.28	0.38	0.83	1.08
Eastern	0.35	0.40	2.21	2.04
Central	0.30	0.58	2.25	2.05
Kordofan	0.41	1.23	5.29	2.30
Darfur	0.41	1.23	5.29	2.30
Khartoum	0.48	0.26	0.93	0.94
Equatorial	0.48	1.40	0.88	0.27
Bahar El Ghazal	0.88	1.36	1.00	0.27
Upper Nile	0.58	0.66	0.76	0.18

Traditional fuels predominate in rural areas; almost all biomass energy is consumed in the household sector for heating, cleaning and cooking needs of rural people. Especially in the villages (households on the high plateau) the preparation of three stone fires is very attractive to the villagers. In this method, food and plant residues are put in a large boiler with water and cooked on a traditional stove at the outside the house for animal feed, because cooked food and plant residues are cheaper than flour and bran. Nevertheless, this method consumes much more fuelwood than the cooking on the stoves method. On the other hand, wood is the most practical fuel for serving a large number of people because the size of the batch of food is only limited by the volume of the pot and not by the size of the stove's burner. Fuelwood is also convenient for cooking of the meal of meat as a cutlet meatball and meat roasted on a revolving vertical spit.

Currently the 'non-commercial' fuels wood, crop residues and animal dung are used in large amounts in the rural areas of Sudan, principally for heating and cooking; the method of use is highly inefficient.

Table 10. National biomass consumption (10^3 kg) [6]

Years	Wood	Charcoal	Other biomass	Total
1980	1969514	1272917	376577	3619008
1983	2024094	1308193	387013	3719300
1985	2080187	1344447	397738	3822372
1987	2137835	1381705	408761	3928301
1990	2197001	1419944	420074	4037019
1993	2255621	1457831	431282	4144734
1995	2315933	1496812	442814	4255559
1997	2377989	1536919	454679	4369587
1998	2623925	1695870	501703	4821498
1999	2689635	1738339	514267	4942241
2000	2765240	1787203	528723	5081166

Table 11. Sustainability matrixes

Power categories	1*	2*	3*	4*	5*	6*	7*	8*	9*	Index
Conventional coal fired stream plant	3	1	1	5	1	1	4	3	3	22
Oil fired stream plant	2	1	1	5	3	3	4	3	3	25
Combined cycle gas turbine	2	3	2	4	4	4	4	2	4	29
Micro combined heat and power	2	3	2	4	4	4	3	2	4	29
Nuclear	4	4	3	5	4	4	3	2	3	32
Hydropower	5	5	5	3	5	5	5	4	2	39
Tidal power	5	5	5	2	5	5	5	4	2	38
Onshore wind	5	5	5	2	5	5	4	4	3	38
Offshore wind	5	5	5	2	5	5	3	4	4	38
Land-fill gases	3	5	3	1	3	4	4	3	2	28
Municipal incineration	5	5	4	3	4	4	4	3	4	36
Biomass, field and forest crops Plus waste straw	5	5	4	3	4	4	4	3	4	36
Import	1	1	5	1	5	5	5	5	5	33
Hydro pumped storage	-	-	5	5	5	5	5	5	2	32
Electrochemical storage	-	-	4	4	4	4	4	4	5	29
Diesel	2	1	1	1	4	5	3	4	4	25

1* fuel availability
2* price stability of fuel
3* by-product acceptability
4* grid services
5* technological obsolescence
6* knowledge base
7* life of the installation
8* maintenance requirement
9* infrastructure requirements

It is useful to codify all aspects of sustainability, thus ensuring that all factors are taken into account for each and every development proposal. Therefore, with the intention of promoting debate, a sustainability matrix is presented (Table 11). The following considerations are proposed:

(1) Long-term availability of the energy source or fuel.
(2) Price stability of energy source or fuel.
(3) Acceptability or otherwise of by-products of the generation process.
(4) Grid services, particularly controllability of real and reactive power output.
(5) Technological stability, likelihood of rapid technical obsolescence.
(6) Knowledge base of applying the technology.
(7) Life of the installation – a dam may last more than 100 years, but a gas turbine probably will not.
(8) Maintenance requirement of the plant.

3.2. Waste Policy in Context

In terms of solid waste management policy, Sudan has changed drastically in the past ten years from waste production society to 'material-cycle society'. In addition to national legislation, municipalities are legally obliged to develop a plan for handling the municipal solid waste (MSW) generated in administrative areas. Such plans contain:

- Estimates of future waste volume.
- Measures to reduce waste.
- Measures to encourage source separation.
- A framework for solid waste disposal and the construction and management of solid waste management facilities.

Alternative fuels were defined as methanol, ethanol, natural gas, propane, hydrogen, in addition to biological material and electricity. The fuel pathways currently under development for alcohol fuels are shown in Figure 2.

3.3. Energy from Agricultural Biomass

The production of agricultural biomass and its exploitation for energy purposes can contribute to alleviate several problems, such as the dependence on import of energy products, the production of food surpluses, the pollution provoked by the use of fossil fuels, the abandonment of land by farmers and the connected urbanisation. Biomass is not at the moment competitive with mineral oil, but, taking into account also indirect costs and giving a value to the aforementioned advantages, public authorities at national and international level can spur its production and use by incentives of different nature. The biomass directly produced by cultivation can be transformed by different processes into gaseous, liquid or solid fuels (Table 12). The main advantages are related to energy, agriculture and environment problems, are foreseeable both at Sudan level and at worldwide level and can be summarised as follows:

- Reduction of dependence on import of energy and related products, and reduction of environmental impact of energy production (greenhouse effect, air pollution, waste degradation).
- Substitution of food crops and reduction of food surpluses and of related economic burdens, and development of new know-how and production of technological innovation.
- Utilisation of marginal lands and of set aside lands and reduction of related socio-economic and environmental problems (soil erosion, urbanisation, landscape deterioration, etc.).

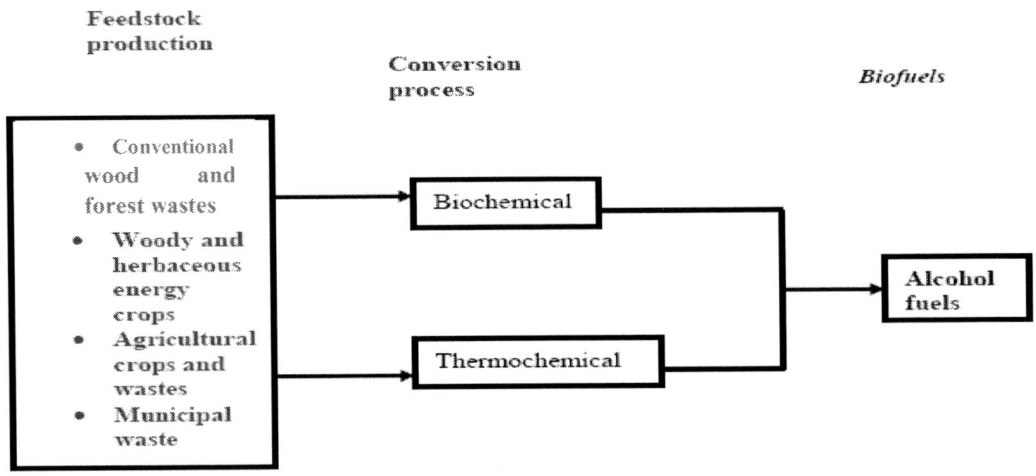

Figure 2. Biofuel pathways for renewable alcohol fuels

Table 12. Biomass conversions to energy

Feedstock	Crops	Conversion process	End product
Wood-cellulosic biomass	Short rotation forest (poplar, willow), plant species (sorghum, mischantus, etc.), fibre-crops (cynara, kenaf, etc.)	Direct combustion Gasification Pyrolysis	Heat Methane Hydrogen Oil
Vegetable oils	Oleaginous crops (rapeseed, soybean, sunflower, etc.)	Direct combustion Esterification	Heat Biodiesel
Sugar/starch	Cereals, root and tuber crops, grape, topinambour, etc.	Fermentation	Ethanol

The whole process of production of methyl or ethyl esters (biodiesel) is summarised in Figure 3. Gasification is based on the formation of a fuel gas (mostly CO and H_2) by partially oxidising raw solid fuel at high temperatures in the presence of steam or air. The technology can use wood chips, groundnut shells, sugar cane bagasse, and other similar fuels to generate capacities from 3 kW to 100 kW. Three types of gasifier designs (Figure 4) have been developed to make use of the diversity of fuel inputs and to meet the requirements of the product gas output (degree of cleanliness, composition, heating value, etc.).

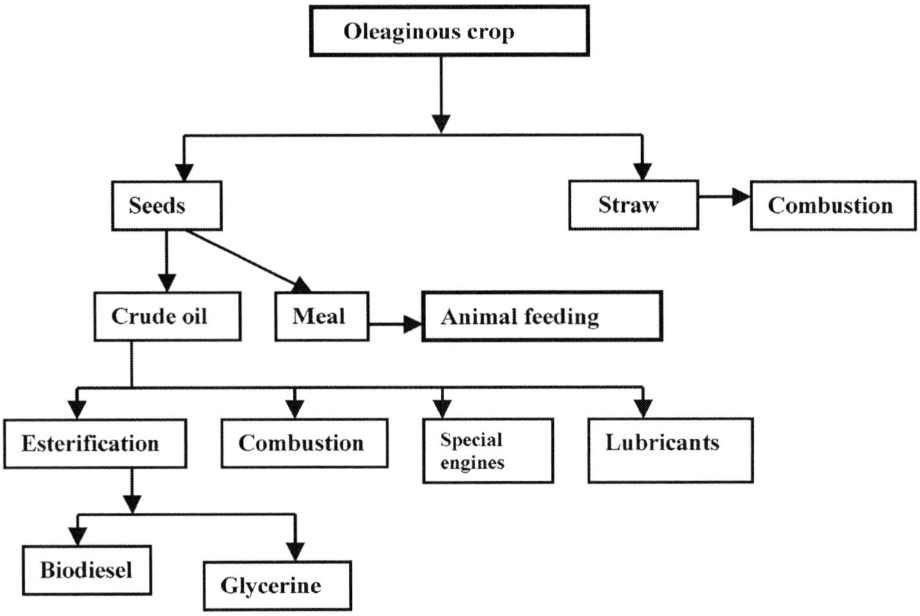

Figure 3. Flow chart of biodiesel production

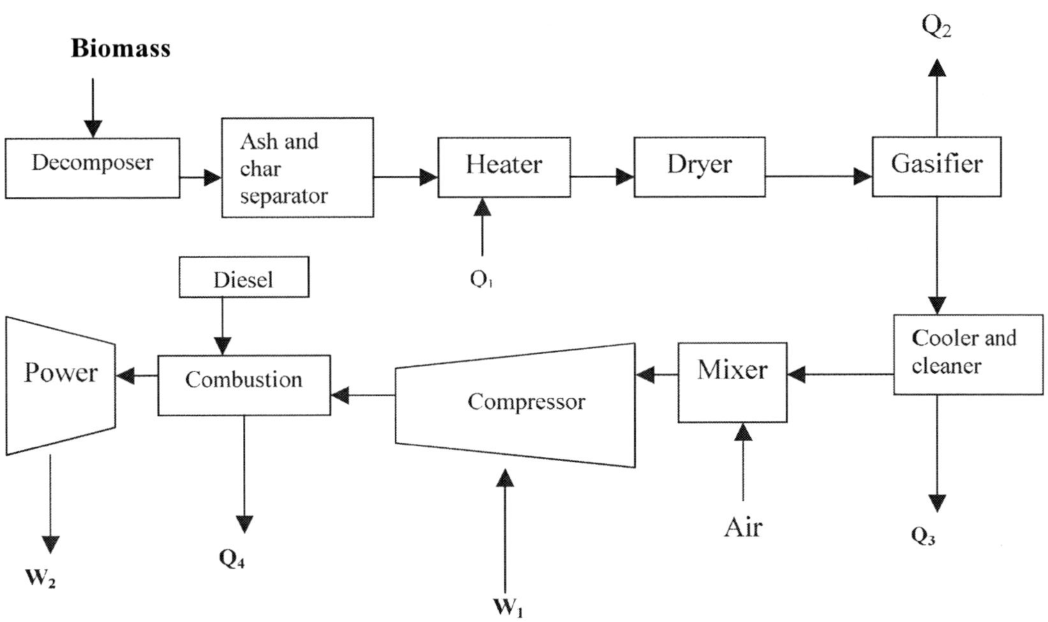

Figure 4. Advanced biomass power with diesel engine

In a study [7] individuated on the basis of botanical, genetical, physiological, biochemical, agronomical and technological knowledge reported some 150 species that are potentially exploitable (Table 13).

Table 13. Plant species potentially exploitable for production of agricultural biomass for energy or industrial utilisations [7]

Groups of plants	Number of species
Plants cultivated for food purposes that can be reconverted to new uses	9
Plants cultivated in the past, but not in culture any more	46
Plants cultivated in other world areas	46
Wild species, both indigenous and exotic	47
Total	148
Plant product	Number of species
Biomass	8
Sugars and polysaccharides	38
Cellulose	17
Hydrocarbons	3
Polymeric hydrocarbons	5
Gums and resins	12
Tannins and phenolic compounds	3
Waxes	7
Vegetable oils	38
Total	131

The production of energy from agricultural biomass seems a realistic option for the future. Some solutions are technically feasible in short term. In some cases environmental, social and politic considerations justify their immediate exploitation even if economic evaluations of merely direct production costs do not justify at the moment of investments. Biomass is a renewable energy source, which can be converted into liquid fuels and/or gas fuels with different technologies available today. Ethanol production via fermentation, extraction and extractive-distillation is one such method and has been practiced for long time in most developing countries with agricultural surplus. However, the intensive research and development activities are still needed.

Data on Sudan energy showed the domination of biomass, which represents more than 87% of the Sudan energy resources, and the household sector consumes more than 92% of the total biomass. Thus the Sudanese energy institutions adopted a strategy focusing on the promotion of conservation measures and introduction of alternative techniques such as direct briquetting and charring oil from the agricultural residues and aquatic weeds. Water hyacinth (Eichhornia Crassipes Solms) is a free-floating aquatic plant, which causes much trouble for navigation in Sudan particularly along White Nile from Kosti up to Malakal (200,000 tons of wet water hyacinth annually). The water hyacinth as a wet substrate represents a promising candidate for biogas production by an aerobic fermentation and has been practiced since mid seventies in Sudan.

Until, recently Sudan was an energy importing country and the energy requirements has been supplied through imports that have caused financial problems. Because of the economical problems in Sudan today, the Sudanese energy policy should be concentrated on assurance of energy supply, reliability, domestic sufficiency, in time, in economic terms, and renewability. Therefore as a renewable energy source, biomass (especially fuelwood) seems interesting because its share of the total energy production at 87% is high and the techniques for converting it to useful energy are simple. On the other hand, biomass may, however, see a great expanded use in response to the environmental problems caused by fossil fuel use in the country. Biomass has been proposed to have a central role to play in future, as a sustainable energy scenario. For this to become a reality several real problems need to be overcome. In Sudan as in other developing countries modernisation of biomass energy provision is an urgent necessity for the sake of human health, protection of the environment, and climate change abatement. Given sufficient recognition, resources and research, biomass could become one of the environmentally friendly fuels of the future in Sudan.

4. Geothermal Energy

Ground source heat pumps (GSHPs) are receiving increasing interest, because of their potential to reduce primary energy consumption and thus reduce emissions of greenhouse gases (GHGs). Different types of heat, i.e., solar energy, waste heat, surplus of heat from co-generation power plants, recovered heat from air conditioning systems, can be stored in the ground in summer and used for heating purposes in winter time. Low and high temperature thermal energy storage can be used in these systems. Low temperature ground storage must be coupled with a heat pump to meet heating requirements. The ground source equipment can operate with lower cost than the electric/gas equipment. The variable-speed heat pump would seem to be the least desirable unit for a utility to advocate since it reduces energy but increases summer demand and has little impact on reducing winter demand. It would appear to be a logical unit to rebate for a winter or summer peaking utility. Energy savings are sufficiently enough to warrant added investments in the \$600 to \$900/ton range, especially if costs can be assumed in a 15 or 30-year mortgage.

4.1. Heat Pumps

Heat pumps function by moving (or pumping) heat from one place to another. Like a standard air-conditioner, a heat pump takes heat from inside a building and dumps it outside the building. The difference is that a heat pump can be reversed to take heat from a heat source outside and pump it inside. Heat pumps use electricity to operate pumps that alternately evaporate and condense a refrigerant fluid to move that heat. In the heating mode, heat pumps are far more "efficient" at converting electricity into usable heat because the electricity is used to move heat, not to generate it. The most common types of heat pumps are the heat pump air-source, heat pump-uses outside air as the heat source during the heating season and the heat sink during the air-conditioning season (Figure 5). Ground-source and water-source heat pumps work the same way, except that the heat source/sink is the ground, groundwater, or a body of surface water, such as a lake. For simplicity, water-source heat pumps are often lumped with ground-source heat pumps.

The great majority of heat pumps work on the principle of the vapour compression cycle. The main components in such a heat pump are the compressor, the expansion valve and the two heat exchangers referred to as the evaporator and the condenser. The components are connected to form a closed circuit. In the evaporator the temperature of the refrigerant is kept lower than the temperature of the heat source. This allows heat to flow from the heat source (ground loops, air or loops in water, e.g., rivers etc.) to the refrigerant.

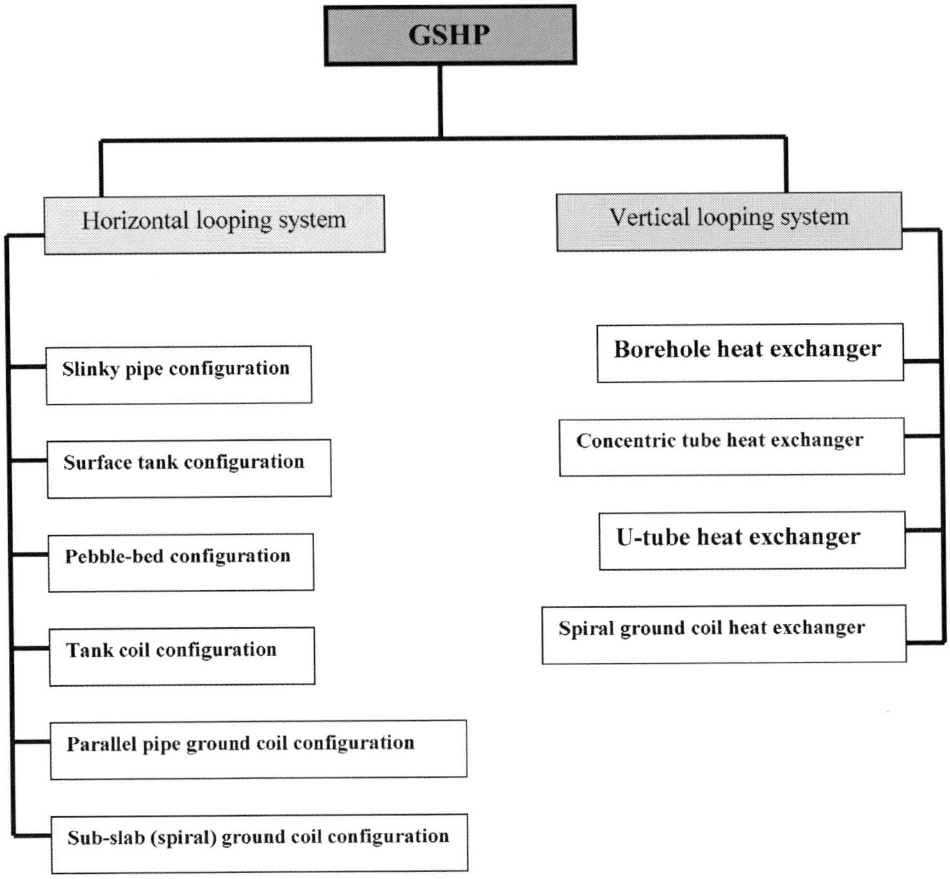

Figure 5. Common types of ground-loop heat exchangers

Ground source heat pumps are the most energy efficient, environmentally clean, and cost effective space conditioning systems available according to the Environmental Protection Agency in the United States of America. Ground source geothermal heating and cooling is a renewable resource, using the earth's energy storage capability. The earth absorbs 47% of the suns energy amounting to 500 times more energy than mankind needs every year.

The efficiency or coefficient of performance (COP) of the GSHPs is significantly higher than that of air-source heat pumps because the heat source is warmer during the heating season and the heat sink is cooler during the cooling season. The GSHPs are environmentally attractive because they deliver so much heat or cooling energy per unit of electricity consumed. The COP is usually 3 or higher. The best GSHPs are more efficient than high-efficiency gas combustion, even when the source efficiency of the electricity is taken into account.

The GSHPs are generally most appropriate for residential and small commercial buildings, such as small-town post offices. In residential and small (skin-dominated) commercial buildings, GSHPs make the most sense in mixed climates with significant heating and cooling loads because the high-cost heat pump replaces both the heating and air-conditioning system. Because GSHPs are expensive to install in residential and small commercial buildings, it sometimes makes better economic sense to invest in energy efficiency measures that significantly reduce heating and cooling loads, and then install less expensive heating and cooling equipment. The savings on equipment cost may be able to pay for most of the envelope improvements. If a GSHP is to be used, planning the site work and project scheduling needed so carefully that the ground loop can be installed with minimum site disturbance or in an area that will be covered by a parking lot or driveway.

The GSHPs are generally classified according to the type of loop used to exchange heat with the heat source/sink. Most common are closed-loop horizontal and closed-loop vertical systems. Using a body of water as the heat source/sink is very effective, but seldom available as an option. Open-loop systems are less common than closed-loop systems due to performance problems (if detritus gets into the heat pump) and risk of contaminating the water source or, in the case of well water, inadequately recharging the aquifer. The GSHPs are complex. Basically, water or a nontoxic antifreeze-water mix is circulated through buried polyethylene or polybutylene piping. This water is then pumped through one of two heat exchangers in the heat pump. When used in the heating mode, this circulating water is pumped through the cold heat exchanger, where its heat is absorbed by evaporation of the refrigerant. The refrigerant is then pumped to the warm heat exchanger, where the refrigerant is condensed, releasing heat in the process. This sequence is reversed for operation in the cooling mode.

Direct-exchange GSHPs use copper ground-loop coils that are charged with refrigerant. This ground loop thus serves as one of the two heat exchangers in the heat pump. The overall efficiency is higher because one of the two separate heat exchangers is eliminated, but the risk of releasing the ozone-depleting refrigerant into the environment is greater if employed. Direct-exchange systems have a small market share.

4.2. Heat Pump Principles

Heat flows naturally from a higher source to a lower temperature source. Heat pumps, however, are able to force the heat flow in the reverse direction, using a relatively small amount of high quality drive energy (electricity, fuel, or high-temperature waste heat). Heat pumps can transfer heat from natural heat sources such as the air, ground or water, to a building. By reversing the heat pump it can also be used for cooling. Heat is transferred in the opposite direction, from the application that is cooled, to surroundings at a higher temperature.

The great majority of heat pumps work on the principle of the vapour compression cycle. The main components of a heat pump are the compressor, the expansion valve and the two heat exchangers referred to as the evaporator and the condenser. The components are connected to form a closed circuit, as shown in Figure 6. A volatile liquid, known as the working fluid or refrigerant, circulates through the four components. The technical and economic performance of a heat pump is closely related to the characteristics of the heat source. An ideal heat source for heat pumps in buildings has a high and stable temperature during the heating season, is abundantly available, is not corrosive or polluted, has favourable thermophysical properties, and its utilisation requires low investment and operational costs. In most cases, however, the availability of the heat source is the key factor determining its use.

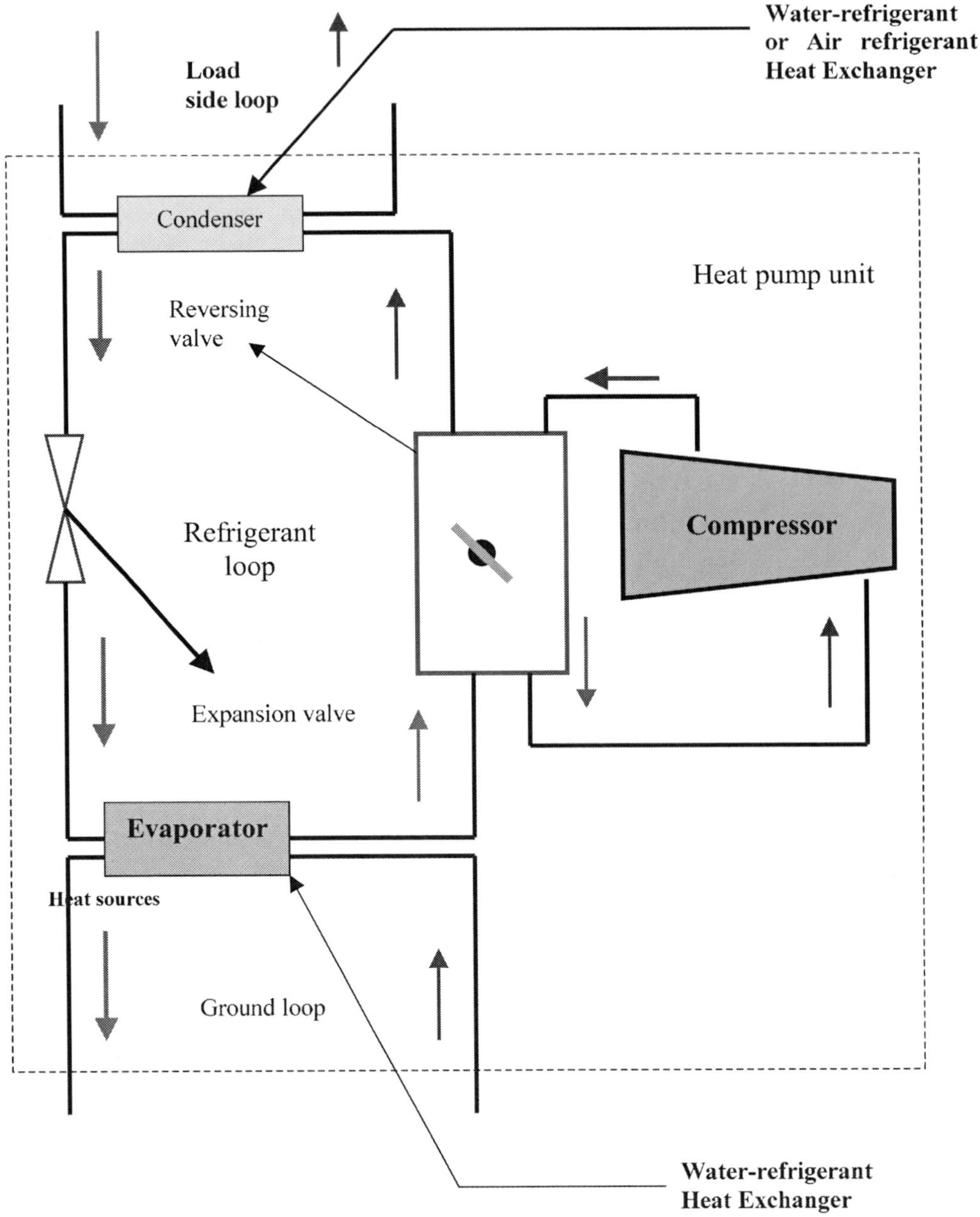

Figure 6. Schematic of GSHP system (heating mode operation)

In order to transport heat, external energy is needed to drive the heat pump. Theoretically, the total heat delivered by the heat pump is equal to the heat extracted from the heat source, plus the amount of drive energy supplied. Electrical powered heat pumps, for heating buildings, typically supply 100 kWh of heat with just 20-40 kWh of electricity. Because heat pumps consume less energy than conventional heating systems, their use will help to reduce the harmful emissions of carbon dioxide, sulpher dioxide and nitrogen oxides. However, the overall environmental impact of electric heat pumps depends very much on how the electricity is produced. Heat pumps driven by electricity generated by hydropower, wind power, photovoltaics or other renewable sources will reduce emissions more significantly than if the electricity is generated by coal, oil or gas-fired power plants.

The technical and economic performance of a heat pump is closely related to the characteristics of the heat source. An ideal heat source for heat pumps in buildings should has a high and stable temperature during the heating season, abundantly available, not corrosive or polluted, and has favourable thermophysical properties, and its utilisation should require low investment and operational costs. In most cases, however, the availability of the heat source is the key factor determining its use. Table 14 presents commonly used heat sources. Ambient and exhaust air, soil and ground water are practical heat sources for small heat pump systems, while sea/lake/river water, rock (geothermal) and waste water are used for large heat pump systems.

Table 14. Heat sources temperatures

Heat source	Temperature range (°C)
Ambient air	-10 - 15
Exhaust air	15 - 25
Ground water	4 - 10
Lake water	0 - 10
River water	0 - 10
Sea water	3 - 8
Rock	0 - 5
Ground	0 - 10

Ground water is available with stable temperatures (4-10°C) in many regions. Open or closed systems are used to tap into this heat source. In open systems the ground water is pumped up, cooled and then reinjected in a separate well or returned to surface water. Open systems should be carefully designed to avoid problems such as freezing, corrosion and fouling. Closed systems can either be direct expansion systems, with the working fluid evaporating in underground heat exchanger pipes, or brine loop systems. Due to the extra internal temperature difference, heat pump brine systems generally have a lower performance, but are easier to maintain. A major disadvantage of ground water heat pumps is the cost of installing the heat source. Additionally, local regulations may impose severe constraints regarding interference with the water table and the possibility of soil pollution.

District heating plays a very important role in transition economies. District heating systems in transition economies often face financial, technical or managerial problems largely created by an inadequate policy framework. These challenges include lack of customer focus, low efficiency, excess capacity, corruption and an uneven playing field. Figure 7 shows how these challenges are interrelated and create a vicious circle that undermines finances and competitiveness of district heating companies, jeopardising their long-term sustainability. Exhaust (ventilation) air is a common heat source for heat pumps in residential and commercial buildings. The heat pump recovers heat from the ventilation air, and provides water and/or space heating

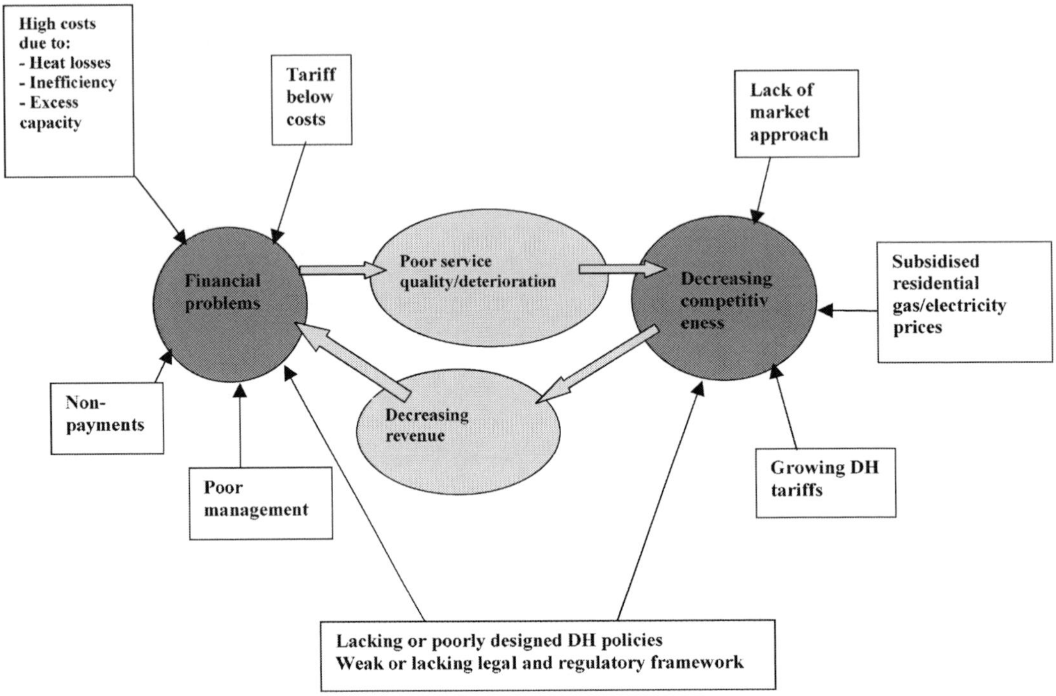

Figure 7. Key challenges for district heating systems in transition economies

4.3. Refrigeration

The refrigeration industry has had to face environmental challenges. Regulation and behaviour related to the ozone layer differ from one country to another. Problems related to global warming are likely to cause similar problems in very different ways. These two kinds of problems may lead company managers to change/modernise equipment. These environmental factors must not hide other important development factors such as those related to hygiene, organoleptic quality, control methods and equipment packaging, etc. All equipment change and investment projects must take all of them into consideration.

The more developed a country is the more widely refrigeration is used. Food preservation is still the main use of refrigeration, followed by air conditioning, energy savings and transport (heat pump and liquefied gases), industrial processes and medicine. When consumers demand more and more fresh products, handling such products becomes increasingly difficult: cold stores, display cabinets, refrigerated trucks have to be much more effective. Difficulties are mostly related to:

- Very narrow ranges, in many cases, between temperatures involving microbial or chemical risk and temperatures that cause chilling injury or freezing.
- No or little overlapping between temperature ranges permitted for different products.

- No or little possibility for the consumer or the seller to evaluate the remaining life spans of the product or even to perceive any risk.
- Humidity control, vapour pressure differences have serious consequences, not only on the weight but also on the unit value of products.

The production and use of Chloro-Fluoro-Carbon (CFC) refrigerants have been phased out in developed countries. Hydro-Chloro-Fluoro-Carbon (HCFC) refrigerants are now subject to regulation and are scheduled to be phased out in the twenty-first century. Recent developments in refrigeration technology and environment (climate) control have provided a better quality of life for mankind. As is known, refrigeration technology has played a key role in preserving the quality of food and eliminating waste (spoilage). The refrigerated transport of perishable products provides a critical link in the cold chain between the consumer and the producers, processors, shippers, distributors and retailers. The transport refrigeration industry incorporated mechanical vapour compression mechanisms into the first transport refrigeration units. Therefore, the physical, chemical and thermodynamic properties of these refrigerants had a major impact on the design of compressors and other refrigeration system components. Many refrigerants have excellent thermodynamic properties suitable for use in vapour compression refrigeration systems, but are limited by other properties such as toxicity, flammability and chemical stability within the refrigeration system.

4.4. Water Source Heat Pump

Water source heat pump (WSHP) systems are one of the most efficient, environmentally friendly ways to heat and cool buildings because of their ability to move energy from where it is not needed to where it is needed. High-efficiency, self-contained WSHP units can be placed in virtually any location within the building and connected via a water loop. Heat is added and rejected from the loop using a boiler and a cooling tower, or using geoexchange from natural sources such as the ground, a pond or a well. Each unit responds only to the heating or cooling load of the individual zone it serves. This provides excellent comfort levels for occupants, better control of energy use for building owners and lower seasonal operating costs. Systems are commonly applied to office buildings, hotels, health care facilities, banks, schools, condominiums and apartments. Features include: Lower utility bills, less maintenance, no visible outdoor plant, reduction in emissions, and versatility of system. Generally, when heat pumps are considered for effective use, the following characteristic features of a heat source will be taken into account:

- Availability,
- Steady and relatively high temperature,
- Coherency between source and user,
- Thermal capacity,
- Heat extraction cannot disturb natural energy balance of the environment,
- Cost of heat extraction, transmission, and ecological issues.

The seasonal storage facility can be designed in many different ways [9]. Heat can be stored in the ground (clay, sand), un-fractured rocks and in water. Four fundamental options of long-term thermal energy storage that can be considered include:

- Water tanks and solar ponds;
- Rocks (boreholes in rocks, rock cavern, pit);
- Soil storage (boreholes, ducts in earth, earth coils);
- Aquifers- systems that use ground water flow.

Ground storage systems can be classified in many different ways. One of the most important classifications is in accordance to the temperature of the storage (Figure 8) as follows:

- GSHPs, without artificial charging the soil - temperature about 10°C.
- Low temperature ground storage - temperature < 50°C.
- High temperature ground storage - temperature > 50°C.

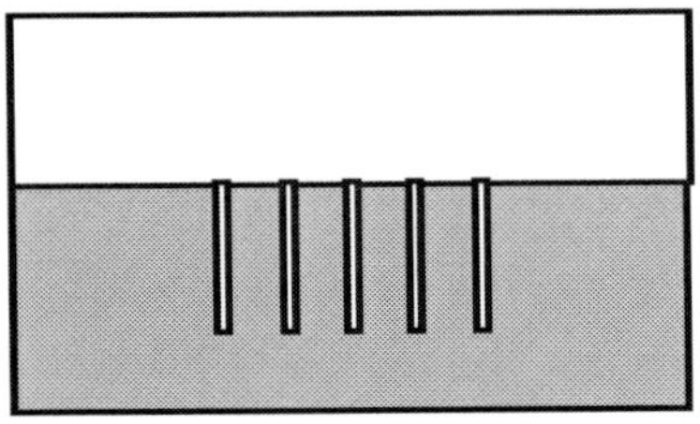

Ground un-separated store vertical

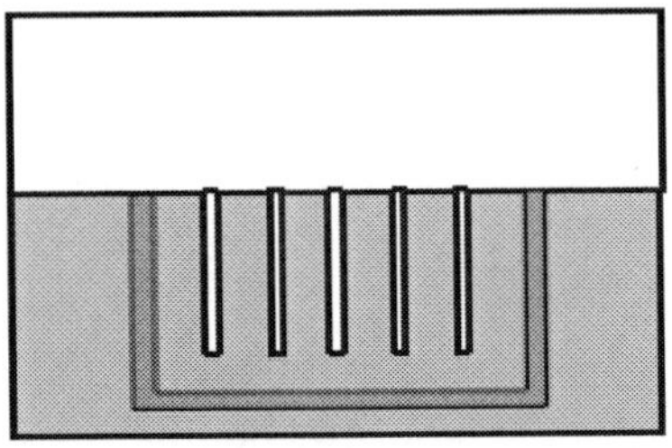

Ground separated store vertical

Ground store horizontal

Figure 8. Typical configurations of ground storage systems

4.5. Geothermal Aquifers

Geothermal aquifers exist where heat from the earth's crust is absorbed by groundwater that collects naturally in the deep porous rocks of certain geological structures. To exploit these aquifers as a source of energy, it is necessary to drill two boreholes: a production borehole to extract the naturally heated water and an injection borehole to dispose the water once that heat has been removed by surface use. It is also possible to use a single hole configuration if the used water is discharged elsewhere – to the sea or to some other convenient sink. Because of the poor thermal conductivity of rock and the low rates of natural fluid recharge, heat is usually extracted at a greater rate than it is replenished from the surrounding rock mass. Geothermal aquifers are not, therefore, "renewable" resources in the strict sense of the word, although they are usually placed in the renewables category. The exploitation of geothermal energy has, to date, concentrated on these hydrothermal resources, most economically from aquifers in countries where seismic activity is high [10].

In Sudan, geothermal have been identified as a potential energy resource [8], and the following sites are expected to have a significant potential:

- Volcanic Jebel Marra area.
- The Red Sea littoral (Suwakin area).
- Volcanic territories.
- Some other remote areas.

Scientific research studies are needed on the above sites for the geothermal energy availability, and then the economic, and social studies can be performed.

4.6. Costs of GSHP

Reductions in GSHP system cost, improvements in installation quality, greater competition, and improved market penetration have occurred primarily in the areas that have been involved with GSHPs for several years. The first cost of GSHPs in areas that do not have established contractors and designers is often prohibitively high. Many potential customers cannot economically justify these relatively high costs. Thus, the GSHP heat pump industry in these areas does not develop sufficiently to support loop and heating, ventilating and air conditioning (HVAC) contractors who will invest in the equipment and training necessary to install GSHPs effectively.

High installation costs have been identified as a major barrier to wider application of GSHPs often referred to as geothermal heat pumps. The primary reason cited for higher cost is the ground loop. Other factors may be high costs of GSHP heat pump units and supplies, interior installation, and limited competition.

The average cost of ground-source heat pump systems ranged from $2,360 per ton (5-ton horizontal) to $3,000 per ton (3-ton vertical) [11]. These figures may vary from place to another depending on many factors. The percentage of the individual components cost are as follows:

Ground loop = 27.2% to 34.1%
Heat pump = 27.1% to 30.1%
Indoor installation = 19.1% to 21.1%
Ductwork = 13.5% to 14.5%
Pumps = 6.1% to 7%

The reasons for these higher costs and lost market opportunities appear to be due to the followings:

1. High cost of ground loops.
2. Higher cost for GSHP heat pump equipment and supplies.
3. Higher cost of HVAC installation.
4. Limited competition.

The way the energy sector operates, especially the electric utility companies, could affect not only the producers but also the customers. Also, the deregulation process may affect the heat pump market in many ways: 1) heat pump economy might be influenced by changes in the energy price structure, and 2) the heat pump market might be stimulated or hindered, depending on changing utility market strategies. So far, in the regulated market, some utilities have clearly supported heat pumps, in line with governmental energy-efficiency programmes (e.g., by offering grants or special electricity tariffs). Only when the market matures and energy prices drop to a stable level will utilities offer incentives such as products/bonuses or energy efficiency services. Nevertheless, the ecological incentives like avoiding GHGs emissions will further support GSHP development. The CO_2 tax in sight is a further (financial) incentive. Of course, there will be considerable differences in this respect from country to country.

5. Solar Energy

Two of the most essential natural resources for all life on the earth and for man's survival are sunlight and water. Sunlight is the driving force behind many of the renewable energy technologies. The worldwide potential for utilising this resource, both directly by means of the solar energy technologies and indirectly by means of biofuels, wind and hydro technologies is vast. During the last decade interest has been refocused on renewable energy sources due to the increasing prices and foreseeable exhaustion of presently used commercial energy sources. In Sudan great attention is given to the utilisation of the available renewable energy resources. Solar energy constitutes the main renewable energy resource in Sudan [12]. There was an attempt to explore the potentialities of solar energy technologies with particular reference to application of solar energy technologies in remote rural areas of Sudan. The main objectives of are reviewing and assessing the current and future share of various solar energy applications in the over all mix of energy in Sudan, as well as identifying potential areas for utilising future technologies and recommending future courses of action to encourage the commercial utilisation of solar energy resources.

5.1. Solar Energy Applications

Sudan enjoys bright sunshine and dry weather most of the year. The utilisation of solar energy in Sudan pertains to be the use of the followings:

5.1.1. Solar cookers

Solar cookers are made from wood and glass. Different types of solar cookers (locally manufactured) have been designed, and disseminated in Sudan during 1970s and 1980s. Now solar cookers are commercially available in the local markets [13].

5.1.2. Industrial solar water heaters (SWHs)

Different systems designs have been developed, tested and evaluated in the country. The best one is the solar water heaters on the roofs of the buildings [13]. A typical family size SWH unit consists of 1-1.44 m^2 of solar collectors' areas and a 1 m^3 hot storage tank. Due to high and reliable solar irradiance of about 5.5 kWh/m^2 day a typical Sudanese SWH with 1.44 m^2 net area and 25% average system efficiency over the life time has the potential to produce around 150 l of hot water at 45°C per day for about 340 sunny days per year. However, several technical and non-technical factors that may influence this application may include the followings: (i) Selection and treatment of materials used in the manufacturing of SWH. This situation drastically reduces the economic viability and performance of SWH systems over their lifetime. (ii) Space availability on roofs especially in high-rise buildings. (iii) Lack of public awareness of the benefits of using SWH systems, and of the maintenance and monitoring procedures, and (iv) The unavailability of credit schemes, especially for the low-income groups, in order to have a SWH system with soft payments.

Currently the Energy Research Institute (ERI) has a SWH laboratory for simulation of solar radiation, quality control, calibration and determination of performance behaviour of SWH. This will assure the quality and reliability of SWH and protect both the consumer and manufacturer.

5.1.3. Solar stills

Various versions of solar stills have been constructed and assessed during 1970s. Solar stills are suitable for production of distilled water for use in laboratories, medical purposes, charging and topping batteries, supplying drinking water to small communities in isolated sunny areas as well as local markets. Solar stills for charging and topping batteries are the one with best economical potential [13].

5.1.4. Solar dryers for peanut crops

Several solar dryers have been developed, tested and proved good in Sudan by ERI. Though these are not used commercially, the technical know-how is currently available.

5.1.5. PV solar lighting systems.

It is essential this application mainly used for schools for night classes, mosques, Quranic schools (*Khalwas*), and household. Currently nearly 1000 units (40 W-1.5 kW) are in operation [13].

5.1.6. PV solar water pumps

While Sudan has a high potential in terms of solar radiation, the utilisation of PVs does not depend only on solar radiation, but also on many other factors. Solar pumps are most cost effective for low power requirement (up to 5 kW) in remote places. Applications include domestic and livestock drinking water supplies, for which the demand is constant throughout the year, and also for irrigation. The suitability of solar pumping for irrigation is uncertain because the demand may vary greatly with the season.

Solar systems may be able to provide trickle irrigation for fruit farming, but are not suitable for the large volumes of water that a needed for wheat growing for example. More than 100 pilot stations employing PV generators for pumping water for drinking purposes, and also for irrigation of small-scale farms are currently operational in Sudan. Solar PV water pumping has been promoted successfully in Kordofan state of Sudan.

5.1.7. PV solar refrigerators

Solar powered refrigerators are usually used for storing medicines in rural remote areas. More than 200 units are installed in Sudan since 1980.

5.1.8. PV communication systems

More than 20 pilot stations employed PV generators for communication purposes, e.g., radios and trains signals (at main stations, e.g., Khartoum, Atbara, Port Sudan, etc.) (50 W-250 W) are in operation [13].

5.2. The Future

There is a considerable solar energy resource in Sudan, which could be of strategic importance in substituting the oil, electricity, wood and charcoal; as well as assisting in rural development; and in improving the quality of life in rural areas. Tapping of solar energy for rural development is, therefore, expected to emerge as an important renewable source of energy in Sudan. There are many other factors, which support this outlook. Among them is the remoteness of rural areas, scarcity of water, abundant sunshine and simplicity of the routine maintenance of solar devices. Above all, solar technologies are already well-established in the country, the potential for which has been demonstrated with a wide range of applications over the past 20 years. These results show that Sudan needs to begin to rely more on solar energy in order to reduce the dependence on imported expensive sources of energy. Therefore, the government activities on solar energy should fall into the following general categories: (1) Direct funding of research, development and demonstration projects. (2) Development of institutional framework to facilitate the use of solar energy (e.g., legislation, regulation), and (3) Measures to improve technical performance and economic Competitives of solar energy technologies (e.g., through proper standards and incentives).

5.3. Greenhouses Applications

Greenhouse cultivation is one of the most absorbing and rewarding forms of gardening for anyone who enjoys growing plants. The enthusiastic gardener can adapt the greenhouse climate to suit a particular group of plants, or raise flowers, fruit and vegetables out of their natural season. The greenhouse can also be used as an essential garden tool, enabling the keen amateur to expand the scope of plants grown in the garden, as well as save money by raising their own plants and vegetables. There was a decline in large private greenhouses during the two world wars due to a shortage of materials for their construction and fuel to heat them. However, in the 1950s mass-produced, small greenhouses became widely available at affordable prices and were used mainly for raising plants [14]. Also, in recent years, the popularity of conservatories attached to the house has soared. Modern double-glazing panels can provide as much insulation as a brick wall to create a comfortable living space, as well as provide an ideal environment in which to grow and display tender plants.

The comfort in a greenhouse depends on many environmental parameters. These include temperature, relative humidity, air quality and lighting. Although greenhouse and conservatory originally both meant a place to house or conserve greens (variegated hollies, cirrus, myrtles and oleanders), a greenhouse today implies a place in which plants are raised while conservatory usually describes a glazed room where plants may or may not play a significant role. Indeed, a greenhouse can be used for so many different purposes. It is, therefore, difficult to decide how to group the information about the plants that can be grown inside a greenhouse.

Throughout the world urban areas have increased in size during recent decades. About 50% of the world's population and approximately 76% in the more developed countries are urban dwellers [14]. Even though there is an evidence to suggest that in many 'advanced' industrialised countries there has been a reversal in the rural-to-urban shift of populations. Virtually all population growth expected between 2000 and 2030 will be concentrated in urban areas of the world. With an expected annual growth of 1.8%, the world's urban population will double in 38 years [14]. This represents a serious contributing to the potential problem of maintaining the required food supply. Inappropriate land use and management, often driven by intensification resulting from high population pressure and market forces, is also a threat to food availability for domestic, livestock and wildlife use. Conversion to cropland and urban-industrial establishments is threatening their integrity. Improved productivity of peri-urban agriculture can, therefore, make a very large contribution to meeting food security needs of cities as well as providing income to the peri-urban farmers. Hence, greenhouses agriculture can become an engine of pro-poor 'trickle-up' growth because of the synergistic effects of agricultural growth such as [14]:

- Increased productivity and increases wealth.

- Intensification by small farmers raises the demand for wage labour more than by larger farmers.
- Intensification drives rural non-farm enterprise and employment.
- Alleviation of rural and peri-urban poverty is likely to have a knock-on decrease of urban poverty.

Despite arguments for continued large-scale collective schemes there is now an increasingly compelling argument in favour of individual technologies for the development of controlled greenhouses. The main points constituting this argument are summarised by [15] as follows:

- Individual technologies enable the poorest of the poor to engage in intensified agricultural production and to reduce their vulnerability.
- Development is encouraged where it is needed most and reaches many more poor households more quickly and at a lower cost.
- Farmer-controlled greenhouses enable farmers to avoid the difficulties of joint management.

Such development brings the following challenges [15]:

- The need to provide farmers with ready access to these individual technologies, repair services and technical assistance.
- Access to markets with worthwhile commodity prices, so that sufficient profitability is realised.
- This type of technology could be a solution to food security problems. For example, in greenhouses, advances in biotechnology like the genetic engineering, tissue culture and market-aided selection have the potential to be applied for raising yields, reducing pesticide excesses and increasing the nutrient value of basic foods.

However, the overall goal is to improve the cities in accordance with the Brundtland Report [16] and the investigation into how urban green could be protected. Indeed, greenhouses can improve the urban environment in multitude of ways. They shape the character of the town and its neighbourhoods, provide places for outdoor recreation, and have important environmental functions such as mitigating the heat island effect, reduce surface water runoff, and creating habitats for wildlife. Following analysis of social, cultural and ecological values of urban green, six criteria in order to evaluate the role of green urban in towns and cities were prescribed [16]. These are as follows:

- Recreation, everyday life and public health.
- Maintenance of biodiversity - preserving diversity within species, between species, ecosystems, and of landscape types in the surrounding countryside.
- City structure - as an important element of urban structure and urban life.
- Cultural identity - enhancing awareness of the history of the city and its cultural traditions.

- Environmental quality of the urban sites - improvement of the local climate, air quality and noise reduction.
- Biological solutions to technical problems in urban areas - establishing close links between technical infrastructure and green-spaces of a city.

The main reasons why it is vital for greenhouses planners and designers to develop a better understanding of greenhouses in high-density housing can be summarised as follows [16]:

- Pressures to return to a higher density form of housing.
- The requirement to provide more sustainable food.
- The urgent need to regenerate the existing, and often decaying, houses built in the higher density, high-rise form, much of which is now suffering from technical problems.

The connection between technical change, economic policies and the environment is of primary importance as observed by most governments in developing countries, whose attempts to attain food self-sufficiency have led them to take the measures that provide incentives for adoption of the Green Revolution Technology [17]. Since, the Green Revolution Technologies were introduced in many countries actively supported by irrigation development, subsidised credit, fertiliser programmes, self-sufficiency was found to be not economically efficient and often adopted for political reasons creating excessive damage to natural resources. Also, many developing countries governments provided direct assistance to farmers to adopt soil conservation measures. They found that high costs of establishment and maintenance and the loss of land to hedgerows are the major constraints to adoption [17]. The soil erosion problem in developing countries reveals that a dynamic view of the problem is necessary to ensure that the important elements of the problem are understood for any remedial measures to be undertaken. The policy environment has, in the past, encouraged unsustainable use of land [17]. In many regions, government policies such as provision of credit facilities, subsidies, price support for certain crops, subsidies for erosion control and tariff protection, have exacerbated the erosion problem. This is because technological approaches to control soil erosion have often been promoted to the exclusion of other effective approaches. However, adoption of conservation measures and the return to conservation depend on the specific agro-ecological conditions, the technologies used and the prices of inputs and outputs of production.

5.3.1. Principle of greenhouse

The greenhouse effect is one result of the differing properties of heat radiation generated at different temperatures. The high temperature sun emits radiation of short wavelength, which can pass through the atmosphere and through glass. Objects inside the greenhouse (or any other building), such as plants, absorb and then re-radiate this heat. Because the objects inside the greenhouse are at a lower temperature than the sun the radiated heat is of longer wavelengths and, hence, cannot re-penetrate the glass. This re-radiated heat is therefore trapped and causes the temperature inside the greenhouse to rise [18]. It has been observed that there is a significant rise of between 2.5 and 15°C in the enclosed room air temperature of a controlled environment greenhouse as reported by various authors [19-24]. This leads to a thermal energy saving for heating a greenhouse and maintaining a favourable environment for crop production during the off-season of up to 75%.

At a certain prescribed depth an imposed temperature equal to a seasonal average ambient temperature, it is possible to identify the main parameters influencing the thermal behaviour of a greenhouse, which lead to a better understanding of the different processes inside the greenhouse. A greenhouse is generally divided into two control volumes, the air inside the covering materials and the semi-infinite space represented by the soil beneath the greenhouse as shown in Figure 9.

The heat balance equation for the cover material is as follows:

$$M_{cover}C_{cover} = dT_{cover}/dt = q_{in\text{-}cover} - q_{out\text{-}cover} \quad (1)$$

Where:

$$q_{in\text{-}cover} = q_{sun\text{-}in\text{-}cover} + q_{conv\text{-}air\text{-}to\text{-}cover} + q_{condens\text{-}air\text{-}to\text{-}cover} \quad (2)$$

and

$$q_{out\text{-}cover} = q_{wind} + q_{thermal\ radiation\text{-}cover} \quad (3)$$

The heat balance equation for the air inside the greenhouse is governed by:

$$m_{air}Cp_{air}\ dT_{air}/dt = q_{out\text{-}air} \quad (4)$$

Where:

$$q_{in\text{-}air} = q_{conv\text{-}soil\text{-}to\text{-}air} + q_{evap\text{-}soil\text{-}to\text{-}air} \quad (5)$$

The one-dimensional heat diffusion equation for semi-infinite soil gives:

$$\rho Cp_{soil} = \partial T_{soil}/\partial t = \alpha_{soil}\ \partial T_{soil}/\partial x^2 \quad (6)$$

The following boundary conditions apply:
At the soil surface a prescribed heat flux resulting from the surface energy balance:

$$q_{condu\text{-}to\text{-}soil} = q_{sun\text{-}to\text{-}soil} - (q_{conv\text{-}soil\text{-}to\text{-}air} + q_{evap\text{-}soil\text{-}to\text{-}air} + q_{thermal\ radiation\text{-}soil}) \quad (7)$$

Where:

Cp_{air}	Specific heat of air (kJ kg^{-1} K^{-1})
M	Air mass flow rate (kgs^{-1})
Q	Volumetric airflow rate (m^3s^{-1})
ρ_a	Air density (kgm^{-3})
T	Air temperature (°C)

Using the above concept with synthetic meteorological data (e.g., solar radiation, ambient temperature, wind speed, relative humidity, ground temperature, thermal and physical properties of the ground) will allow the prediction of energy consumption during a whole season in order to maintain the required inside air conditions for plant growth. Further modifications and introducing the more complex problem of plant growth, will allow programming the greenhouse for specific plants, for a specific types of solar collector.

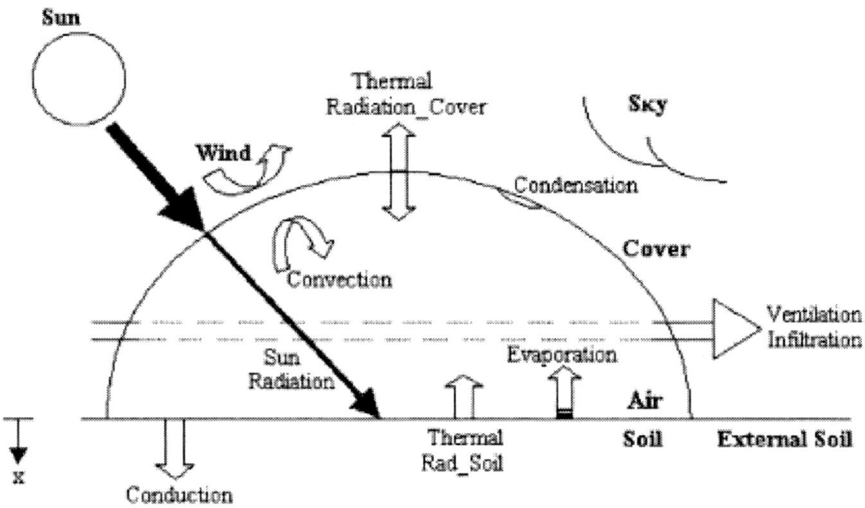

Figure 9. Greenhouse heat and mass transfer

5.3.2. Heating the greenhouse or conservatory

Heating a greenhouse in the winter is a major cost and an early decision as to the minimum winter temperature to maintain will, in turn, determine which plants can be grown. Greenhouse plants are normally classified in terms of their temperature requirements as follows [25]:

FF - Frost Free Only
C - Cool (minimum of 7°C)
W - Warm (minimum of 13°C)
T - Tropical (minimum of 18°C)

There are various methods of heating a greenhouse or a conservatory. These include gas or electric heaters, paraffin heaters and a boiler system with hot-water pipes. Both paraffin and gas heaters give off a lot of moisture making the air more humid than is desirable in winter and increasing condensation. Therefore, a small amount of ventilation must always be kept on to allow air to circulate and to allow harmful gases to escape. On the other hand, if the greenhouse is a long way from the house it may prove impossible or too expensive to run a source of electricity down to it. There is, also, the possibility of power failures to contemplate.

Table 15, which is reproduced from [26], classifies plants according to their climate preferences. The table, also, indicates that there are many plants that can be grown in a humid 'microclimate'.

Table 15. Atmospheric conditions for plants [26]

Types/conditions	Warm conditions and high humidity	Warm conditions and thrive in a dry atmosphere	Cool
Shrubs	Aphelandra, Gardenia	Lantana, Cestrum	Acacia, Lantana
Climbers	Allamanda, Hoya	Plumbago	Jasminum, Plumbago
Flowering pot plants short-term	Capsicum, Celosia	Ipomoea	Solanum, Primula
Flowering pot plants long-term	Begonia, Euphorbia	Fuchsis, Cacti	Clivia, Erica
Foliage pot plants	Ferns, Ficus	Sansevieria	Coleus, Hedera
Bulbous and tuberous plants	Canna	Vallota, Gloriosa	Cyclamen, Lilium
Fruits	Citrus	-	Vitis Vinifera, Prunus Persica

5.3.3. Indoor environment control

Control of greenhouse indoor environmental conditions in the winter and summer months are often relatively straight forward as only heating or cooling are required respectively. There is often a large difference between the required indoor conditions and the outdoor conditions, e.g., temperature, relative humidity. However, during the mid-seasons opportunity often exists to take advantage of phase and value differences between indoor and outdoor environments to enable improvements in energy efficiency of building environmental services operation. Consequentially, opportunity sometimes exists to increase fresh air ventilation rates and reduce indoor air temperatures without the need for mechanical cooling. Simultaneously, the benefit of improved indoor air quality is realised as a result of increased fresh air ventilation rates. It is also essential that cost efficiency is to be taken into account and not be compromised in favour of energy efficiency as this would negatively impact on the overall building performance [27]. Hence, the use of a control system is highly recommended.

Indeed, control system is important to maintain indoor environmental quality within a control volume, i.e., within predetermined ranges for each of the environmental parameters, while considering the best course of action with respect to energy and cost efficiencies. Temperature and relative humidity were assigned upper limits, lower limits, and preferred values to be sought when they were achievable without adversely affecting cost or energy efficiencies.

6. Environment, Climate Change and Global Warming

With the environmental protection posing as the number one global problem, man has no choice but reducing his energy consumption, one way to accomplish this is to resort to passive and low-energy systems to maintain thermal comfort in buildings. The conventional and modern designs of wind towers can successfully be used in the hot arid regions to maintain thermal comfort (with or without the use of ceiling fans) during all hours of the cooling season, or a part of this period. Climatic design is one of the best approaches to reduce the energy cost in buildings. Proper design is the first step of defense against stress of climate. Buildings should be designed according to climate of the site for reducing the need of mechanical heating or cooling hence maximum natural energy can be used for creating pleasant environment inside the built envelope. Technology and industry progress of the last decade diffused electronic and informatics' devices in many human activities and now appear also in building construction. The utilisation and operating opportunities components, increase the reduction of heat losses by varying the thermal insulation, optimise the lighting distribution with louver screens and operate mechanical ventilation for coolness in indoor spaces. In addiction to these parameters the intelligent envelope can act for security control and became an important part of the building domotic revolution. Application of simple passive cooling measure is effective in reducing the cooling load of buildings in hot and humid climates. A 43% reduction can be achieved using a combination of well-established technologies such as glazing, shading, insulation, and natural ventilation. More advanced passive cooling techniques such as roof pond, dynamic insulation, and evaporative water jacket need to be considered more closely. The building sector is a major consumer of both energy and materials worldwide, and the consumption is increasing. Most industrialised countries are becoming more and more dependent on external supplies of conventional energy carriers, i.e., fossil fuels. Energy for heating and cooling can be replaced by new renewable energy sources. New renewable energy sources, however, are usually not economically feasible compared with the traditional carriers. In order to achieve the major changes needed to alleviate the environmental impacts of the building sector, it is necessary to change and develop both the processes in the industry itself, and to build a favourable framework to overcome the present economic, regulatory and institutional barriers.

7. CONCLUSIONS

Until recently, Sudan was an energy importing country and the energy requirements has been supplied through imports that have caused financial problems. Because of the economical problems in Sudan today, the Sudanese energy policy should be concentrated on assurance of energy supply, reliability, domestic sufficiency, in time, in economic terms, and renewability. Therefore as a renewable energy source, biomass (especially fuelwood) seems interesting because its share of the total energy production at 87% is high and the techniques for converting it to useful energy are easily available. On the other hand, biomass may, however, see a great expanded use in response to the environmental problems caused by fossil fuel use in the country. Biomass has been proposed to have a central role to play in future, more sustainable energy scenarios. For this to become a reality several real problems need to be overcome. In Sudan as in other developing countries modernisation of renewable energy provision is an urgent necessity for the sake of human health, protection of the environment, and climate change abatement. Given sufficient recognition, resources and research renewable could become the environmentally friendly fuel of the future.

NOMENCLATURE

Cp_{air}	Specific heat of air (kJ kg^{-1} K^{-1})
M	Air mass flow rate (kgs^{-1})
Q	Volumetric airflow rate (m^3s^{-1})
ρ_a	Air density (kgm^{-3})
T	Air temperature (^oC)
FF	Frost Free Only
C	Cool (minimum of 7^oC)
W	Warm (minimum of 13^oC)
T	Tropical (minimum of 18^oC)

REFERENCES

[1] National Electricity Corporation (NEC). (2006). Annual Report. Khartoum: Sudan.

[2] Omer, A. M. (June 2002). Strengthening community water supply management and striving for improved sanitation in Sudan. *Water International, 27 (2),* 225-232.

[3] World Bank (WB). (1990). Sudan forestry sector review.

[4] Bank of Sudan (BOS). (1997). Annual Report. Khartoum: Sudan.

[5] National Forestry Corporation (NFC). (1986). Sudan Forestry and Wood-fuel Resources Assessment. Khartoum: Sudan.

[6] Omer, A.M. (1996). Renewable energy potential and future prospects in Sudan. *Agriculture Development in Arab World, 3*, 4-13.

[7] Rossi, S., Arnone, S., Lai, A., Lapenta, E., and Sonnino, A. (1990). ENEA's activities for developing new crops for energy and industry. In: Biomass for Energy and Industry (G. Grassi, G. Gosse, G. dos Santos eds.), *Elsevier Applied Science*, *Vol.1,* p.107-113, London and New York.

[8] Omer, A. M. (11-16 September 1994). Renewable energy technology applications in Sudan. In: *Proceedings of the 3rd World Renewable Energy Congress*. *Reading*, UK.

[9] Fridleifsson, I. (1997). Worldwide Geothermal Development and International Cooperation. Proc. *NEDO International Geothermal Symposium Sendai*, 8-15.

[10] Air-Conditioning and Refrigeration Institute (1991). Directory of Certified Applied Air-Conditioning Products. Arlington, VA.

[11] Bose, J. E. (1988). Closed-Loop/Ground-Source Heat Pump Systems. *International Ground-Source Heat Pump Association*. Stillwater, OK.

[12] Omer, A. M. (1998). Sudan energy background; an overview, *Renewable Energy Journal,* Vol.*14,* No. *(1-4)*, p.467-472, Reading, UK.

[13] Energy Research Institute (ERI). (1997). Renewable energy resources potential in Sudan, Khartoum, Sudan.

[14] John, W. (1993). The glasshouse garden. *The Royal Horticultural Society Collection.* UK.

[15] United Nations. (2001). World Urbanisation Prospect: The 1999 Revision. New York. *The United Nations Population Division.*

[16] WCED. (1987). Our common future. New York. Oxford University Press.

[17] Herath, G. (1985). The Green Revolution in Asia: productivity, employment and the role of policies. *Oxford Agrarian Studies, 14,* 52-71.

[18] BRECSU. (1998). Energy use in offices. *Energy Consumption Guide 19.* Watford: Building Research Energy Conservation Support Unit.

[19] Farm Energy Centre. (2000). Helping agriculture and horticulture through technology, *energy efficiency and environmental protection.* Warwickshire.

[20] Randall M. (1992). Environmental Science in Building. Third Edition.

[21] Tiwari, G. N. & Goyal, R. K. (1998). Greenhouse technology. New Delhi: Narosa Publishing House.

[22] Santamouris, M., Balaras, C. A., Dascalaki, E., and Vallindras, M. (1994). Passive solar agricultural greenhouse; a worldwide classification evaluation of technologies and systems used for heating purpose. *Solar Energy, 53(5),* 411-26.

[23] Santamouris, M., Arigirious, A., and Vallindras, M. (1993). Design and operation of a low energy consumption passive greenhouse. *Solar Energy, 52(5),* 371-8.

[24] Mercier, I. (1982). Design and operation of a solar passive greenhouse in the South West France. In: *Proceedings of the International Congress on Energy Conservation of Agriculture and Fishculture.* London.

[25] Anne, S. (1989). A handbook of greenhouse and conservatory plants. Paston Press. London.

[26] Lynn, B. (1993). The pleasure of gardening. ANAYA Publishers Limited. London.

[27] Wu, J., and Boggess, W. (1999). The optimal allocation of conservation funds. *Journal of Environmental Economic Management,* p.38.

[28] Omer, A. M. (2005). Biomass energy potential and future prospect in Sudan. *Renewable and Sustainable Energy Reviews,* 9: 1-27.

In: Energy Costs, International Developments…
Ed: Leszek Kowalczyk and Jakub Piotrowski
ISBN: 978-1-60741-700-2

Chapter 2

Non-Conventional Energy Systems and Environmental Pollution Control

Abdeen Mustafa Omer
17 Juniper Court, Forest Road West, Nottingham NG7 4EU, UK

Abstract

The massive increases in fuel prices over the last years have however, made any scheme not requiring fuel appear to be more attractive and to be worth reinvestigation. In considering the atmosphere and the oceans as energy sources the four main contenders are wind power, wave power, tidal and power from ocean thermal gradients. The renewable energy resources are particularly suited for the provision of rural power supplies and a major advantage is that equipment such as flat plate solar driers, wind machines, etc., can be constructed using local resources and without the advantage results from the feasibility of local maintenance and the general encouragement such local manufacture gives to the build up of small-scale rural based industry. The key factors to reducing and controlling CO_2, which is the major contributor to global warming, are the use of alternative approaches to energy generation and the exploration of how these alternatives are used today and may be used in the future as green energy sources. Even with modest assumptions about the availability of land, comprehensive fuel-wood farming programmes offer significant energy, economic and environmental benefits. These benefits would be dispersed in rural areas where they are greatly needed and can serve as linkages for further rural economic development. Self-renewing resources such as wind, sun, plants and heat from the earth can provide clean abundant energy through the development of renewable technologies. Virtually all regions of the world have renewable resources of one type or another. Research and development investments in the past 25 years in renewable technologies development has lead to important advances in performance and resulting cost effectiveness. Renewable resources currently account for about 9%-10% of the energy consumed in the world; most of this is from hydropower and traditional biomass sources. Wind, solar, biomass and geothermal technologies are cost effective today in an increasing number of markets and are making important steps to broader commercialisation. The present situation is best characterised as one of very rapid growth for wind and solar technologies and of significant promise for biomass and geothermal technologies. Each of the renewable energy technologies is in a different stage of research, development and commercialisation and all have differences in current and future expected costs, current industrial base, resource availability and potential impact on energy supply. This article discusses the potential for such integrated systems in the stationary and portable power market in response to the critical need for a cleaner energy technology. Anticipated patterns of future energy use and consequent environmental impacts (acid precipitation, ozone depletion and the greenhouse effect or global warming) are comprehensively discussed in this paper. Throughout the theme several issues relating to renewable energies, environment and sustainable development are examined from both current and future perspectives.

Keywords: Energy, renewable energy technologies, energy efficiency, environment, sustainable development, global warming, emissions.

1. Introduction

Energy has been a vital input into the economic and social development. However, one third of the world population, living in developing and threshold countries, has no access to electricity. These people mostly live in remote and rural areas with low population density, lacking even the basic infrastructure. Accordingly, utility grid extension is not a cost-effective option and sometimes technically not feasible. Therefore, it is imperative to look for sustainable (i.e., cost-effective, environmentally benign and reliable) sources of energy for the development of these regions. Using locally available renewable energy sources (especially solar irradiation that is characterised by a sufficient availability on a daily basis), which are of high potential in most of these regions offers a strategic solution for their techno-economic development. From the point of view of technology, the design of system technology that meets electrification requirements and fulfils, if necessary, the requirements of integration into alternative current (AC) supply grids, has to be considered.

The modernisation of the system components and their power ranges which allow easy expandability of the supply structure, the standardisation of interfaces and the hybridisation by integration of different energy converters in order to increase the power availability, represent the most important measures from the point of view of system technology. Moreover, the use of renewable energy sources is essentially made easier if the existing reliable AC- technical standards of construction and extension of conventional electricity supply systems are adopted. Therefore, incompatibility cannot be taken as a reason to reduce the dissemination of renewable systems.

In the early years of photovoltaic (PV) history stand-alone as well as grid-connected PV-systems had been built as individual items or unique masterpieces. Until recent years the realisation of PV-systems still was characterised by monolithic system concepts resulting in a costly design and engineering process. Consequently a large number of different PV system components (e.g., inverter) were developed each tailored for the use in the dedicated application with its specific parameters (e.g., input voltage/current).

The exploitation of the energetic potential (solar and wind) for the production of electricity proves to be an adequate solution in isolated regions where the extension of the grid network would be a financial constraint. The use of wind as alternative energy source is increasing and research and development about this clean and unlimited resource is being carried out on various levels.

Likewise, energy savings from the avoidance of air conditioning can be very substantial. Whilst day-lighting strategies need to be integrated with artificial lighting systems in order to become beneficial in terms of energy use, reductions in overall energy consumption levels by employment of a sustained programme of energy consumption strategies and measures would have considerable benefits within the buildings sector. The perception often given however is that rigorous energy conservation as an end in itself imposes a style on building design resulting in a restricted aesthetic solution. It would perhaps be better to support a climate sensitive design approach, which encompassed some elements of the pure conservation strategy together with strategies, which work with the local ambient conditions making use of energy technology systems, such as solar energy, where feasible. In practice, low energy environments are achieved through a combination of measures that include:

- The application of environmental regulations and policy.
- The application of environmental science and best practice.
- Mathematical modelling and simulation.
- Environmental design and engineering.
- Construction and commissioning.
- Management and modifications of environments in use.

In summary, achieving low energy building requires comprehensive strategy that covers; not only building designs, but also considers the environment around them in an integral manner. Major elements for implementing such a strategy are as follows:

1.1. Efficiency Use of Energy

- Climate responsiveness of buildings.
- Good urban planning and architectural design.
- Good house keeping and design practices.
- Passive design and natural ventilation.
- Use landscape as a means of thermal control.
- Energy efficiency lighting.
- Energy efficiency air conditioning.
- Energy efficiency household and office appliances.
- Heat pumps and energy recovery equipment.
- Combined cooling systems.
- Fuel cells development.

1.2. Utilise Renewable Energy

- Photovoltaics.
- Wind energy.
- Small hydros.

- Waste-to-energy.
- Landfill gas.
- Biomass energy and biofuels.

1.3. Reduce Transport Energy

- Reduce the need to travel.
- Reduce the level of car reliance.
- Promote walking and cycling.
- Use efficient public mass transport.
- Alternative sources of energy and fuels.

1.4. Increase Awareness

- Promote awareness and education.
- Encourage good practices and environmentally sound technologies.
- Overcome institutional and economic barriers.
- Stimulate energy efficiency and renewable energy markets.

2. Energy from Waste

Measures to maximise the use of high-efficiency generation plants and on-site renewable energy resources are important for raising the overall level of energy efficiency. The world's view of waste has changed dramatically in recent years and it is now seen as a source to feed the ever-growing demand for energy (Figure 1). The road from the initial concept to the production of the first kilowatt of power is long and has many challenges, not least the need for adequate funding. Scientific research evidence, public awareness and increased levels of participation in environmental campaigning have led to governments' worldwide implementing regulations and legislation. Examples include:

- EU landfill diversion directive.
- Recycling targets.
- Climate change regulations.

The demand for nuclear power generation, wind farms, solar power and so on is now unstoppable and has created a whole new market, though each has its own challenges (Figure 2). The waste collection, transfer and landfill disposal business comprise a mature, slow-growth industry. Economic drivers to developing the waste and renewable energy sector have included:

- Waste disposal and landfill gate fees/landfill tax.

- Penalties/avoidance schemes (e.g., landfill allowance schemes and fines, carbon trading).
- Energy prices.
- Investments subsidies.

When considering the demand and opportunity in today's marketplace these points are prevalent:

- The demand for renewable energy is not going to go away.
- The public feeling is that governments across the world are responsible.
- The pressure caused by diminishing fossil fuel supplies is increasing.
- Investment funds are increasingly available from traditional sources.
- The needs for new technologies that can deliver carbon reduction and waste reduction outcomes are increasingly bankable, which opens up the market for all.

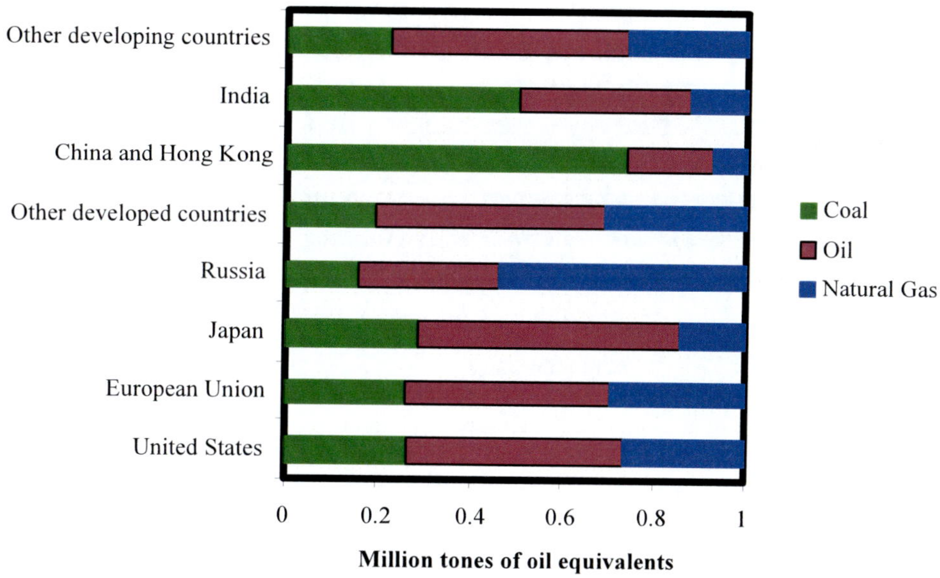

Figure 1. Global fossil fuel consumption

Financial institutions across most global markets are gearing themselves up for the environmental revolution. Within the waste to renewable energy sector, history has shown a hesitancy to invest in projects not supported by four things:

- Adequate independent technology due diligence.
- Security of waste input and power off-take contracts.
- A site with planning permission.
- A reference plant, preferably at scale.

Reviewing the evolution of municipal sewage waste (MSW) management in general, waste collection has tended to progress from incomplete collection through to complete collection and finally to collection with separation into different waste streams. In turn, waste treatment has progressed from ad-hoc decentralised disposal to a strategy more dependent on controlled treatment and disposal, including the use of sanitary landfilling accompanied by waste reduction strategies. In developed countries, this evolution has taken place over a period of about 30-40 years. The standard of liner design and construction standard are summarised in Table 1.

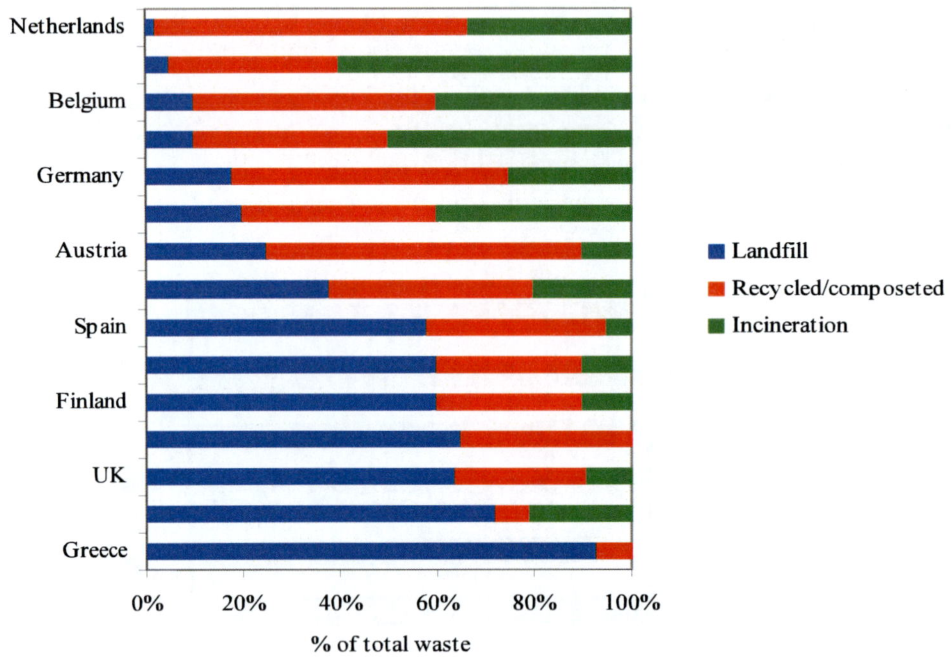

Figure 2. Municipal waste management in the European Union

Table 1. Comparison of basic requirements for bottom liners in MSW sanitary landfills

Liner system requirement	Leachate drainage layer	Geomembrane line	Compacted clay layer
US EPA Standard (40CFR258)	$K > 1x10^{-4}$ ms^{-3} Thickness 0.3 m	Thickness ≥0.75 mm Recommended 1.5 mm HDPE	$K \leq 1x\ 10^{-6}$ ms^{-3} Thickness ≈60 cm
EU Landfill Directive (1999/31/Dec)	Thickness 0.5 m	Not specified. Yet liner thickness should be 100 cm $K \leq 1x\ 10^{-6}$ ms^{-3}	With HDPE liner, thickness of clay layer > 50 cm
German Standard (TASI 1993)	$K > 1x10^{-3}$ ms^{-3} Thickness 0.3 m	Thickness ≥2.5 mm HDPE	$K \leq 5x\ 10^{-10}$ ms^{-3} Thickness 3x25 cm
Chinese Standard (CJJ 113-2007)	$K > 1x10^{-3}$ ms^{-3} Thickness >0.3 m	Thickness ≥1.5 mm HDPE	$K \leq 1x\ 10^{-6}$ ms^{-3} Thickness 75 cm

One of the negative results of growing prosperity worldwide has been an increase in waste generation from year to year. In response, policy-makers and researchers are examining how best to decouple waste growth and economic growth. In both developed and developing countries sanitary landfill sites can be operated in such a way that danger to residents and the environment, from Leachate, odours, fire and explosion is almost entirely eliminated. Table 2 summarised different parameters in waste compaction. Waste professionals use cross wrap machinery for its reliability and efficiency in storage and transport of waste materials.

2.1. Waste Shredding

With the demand for faster and more efficient recycling technologies showing no signs of abating, the market for faster, more efficient shredding equipment is of course on the up. To the man on the street the term shredding most likely brings to mind the transformation of business documents, bank and credit card statements into a bird's nest of paper- a practice now relies on worldwide to prevent fraudsters accessing the personal financial data and sensitive information. It means big business, as shredding of waste is common practice across almost all areas of the waste industry. Far from focusing simply on paper, shredding is a disposal technique for everything from agriculture to household waste and electrical to industrial waste.

The overall trend in today's market tends to be 'shred first and sort later'. Shredding of waste material as a precursor to sorting is useful for two reasons. It reduces the size of the waste, allowing for greater ease of transportation, but perhaps more importantly – at a time when recycling as much materials as effectively as possible is paramount- it allows for more effective sorting afterwards. And logically, effective sorting equals greater opportunity for recycling (Figure 3).

Table 2. The different parameters in waste compaction

Refuse	Item size Organic components Inert substances Slurry
Application technique	Thin layer operation Face operation Pushing distance
Compaction machine	Compactor Operating weight Wheel design
Types of waste disposal site	Pit type waste disposal site Raised refused disposal site Height of the refuse disposal site Refuse load
Weather conditions	Precipitation Temperature

While every customer on the lookout for a shredder is interested in efficiency, one thing that will also attract a potential buyer is energy efficiency. With the large environmental challenges facing the world today, waste industry professionals are increasingly aware of the need to make sure their business are as kind to the environment as possible. The operation of large-scale equipment such as shredding machines naturally uses a large amount of power and machines which can run effectively on a lesser amount have an advantage over their competitors.

Even with modest assumptions about the availability of land, comprehensive fuel-wood farming programmes offer significant energy, economic and environmental benefits. These benefits would be dispersed in rural areas where they are greatly needed and can serve as linkages for further rural economic development. The nations, as a whole would benefit from savings in foreign exchange, improved energy security, and socio-economic improvements. With a nine-fold increase in forest – plantation cover, the nation's resource base would be greatly improved. The international community would benefit from pollution reduction, climate mitigation, and the increased trading opportunities that arise from new income sources. The aim of any modern biomass energy systems must be:

- To maximise yields with minimum inputs.
- Utilisation and selection of adequate plant materials and processes.
- Optimum use of land, water, and fertiliser.
- Create an adequate infrastructure and strong R & D base.

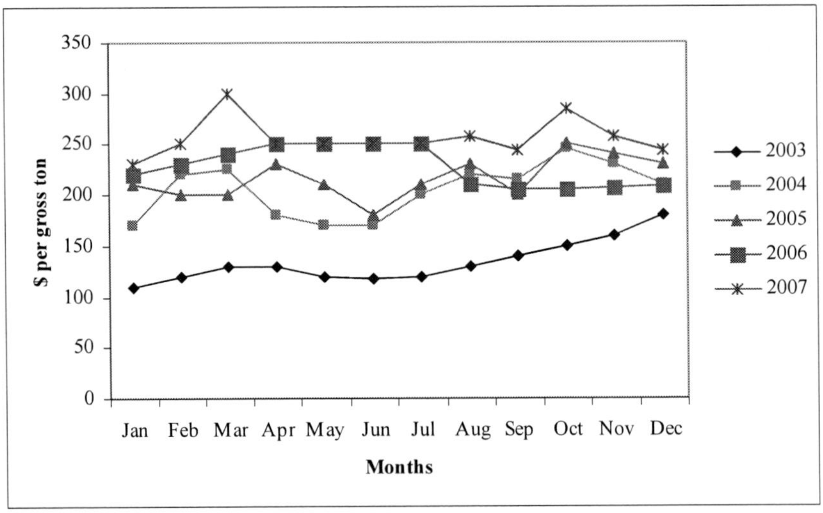

Figure 3. Monthly averages paper scraps

2.2. Biomass CHP

Combined heat and power (CHP) installations are quite common in greenhouses, which grow high-energy, input crops (e.g., salad vegetables, pot plants, etc.). Scientific assumptions for a short-term energy strategy suggest, that the most economically efficient way to replace the thermal plants is to modernise existing power plants to increase their energy efficiency and to improve their environmental performance. However, utilisation of wind power and the conversion of gas-fired CHP plants to biomass would significantly reduce the dependence on imported fossil fuels. Although a lack of generating capacity is forecasted in the long-term, utilisation of the existing renewable energy potential and the huge possibilities for increasing energy efficiency are sufficient to meet future energy demands in the short-term.

A total shift towards a sustainable energy system is a complex and long process, but is one that can be achieved within a period of about 20 years. Implementation will require initial investment, long-term national strategies and action plans. However, the changes will have a number of benefits including: a more stable energy supply than at present and an improvement in the environmental performance of the energy sector, and certain social benefits. A vision used a methodology and calculations based on computer modelling that utilised:

- Data from existing governmental programmes.
- Potential renewable energy sources and energy efficiency improvements.
- Assumptions for future economy growth.
- Information from studies and surveys on the recent situation in the energy sector.

In addition to realising the economic potential identified by the National Energy Savings Programme, a long-term effort leading to a 3% reduction in specific electricity demand per year after 2020 is proposed. This will require: further improvements in building codes, and continued information on energy efficiency. The environmental Non Governmental Organisations (NGOs) are urging the government to adopt sustainable development of the energy sector by:

- Diversifying of primary energy sources to increase the contribution of renewable and local energy resources in the total energy balance.
- Implementing measures for energy efficiency increase at the demand side and in the energy generation sector.

Methane is a primary constituent of landfill gas (LFG) and a potent greenhouse gas (GHG) when released into the atmosphere. Globally, landfills are the third largest anthropogenic emission source, accounting for about 13% of methane emissions or over 818 million tones of carbon dioxide equivalent ($MMTCO_2e$) [1] as shown in Figure 4.

The price of natural gas is set by a number of market and regulatory factors that include:

Supply and demand balance and market fundamentals, weather, pipeline availability and deliverability, storage inventory, new supply sources, prices of other energy alternatives and regulatory issues and uncertainty.

Classic management approaches to risk are well documented and used in many industries. This includes the following four broad approaches to risk:

- Avoidance includes not performing an activity that could carry risk. Avoidance may seem the answer to all risks, but avoiding risks also means losing out on potential gain.

- Mitigation/reduction involves methods that reduce the severity of potential loss.
- Retention/acceptance involves accepting the loss when it occurs. Risk retention is a viable strategy for small risks. All risks that are not avoided or transferred are retained by default.
- Transfer means causing another party to accept the risk, typically by contract.

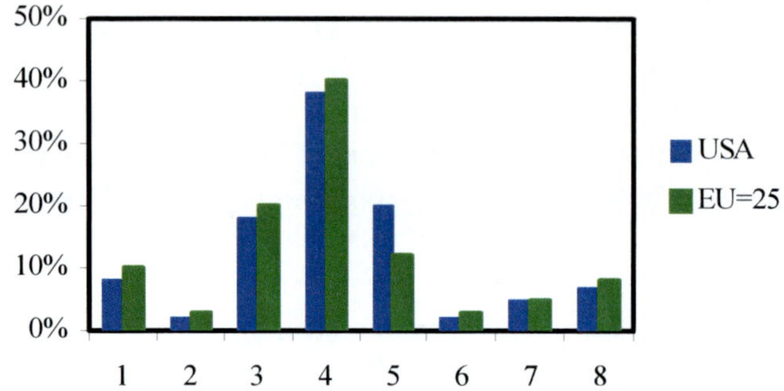

1 Food, 2 Textile, 3 Pulp & paper, 4 Chemicals, 5 Refining, 6 Minerals, 7 Primary metals, and 8 Others

Figure 4. Distribution of industrial CHP capacity in the EU and USA

3. Wind Energy

There are numerous factors that influence the overall prospects for the wind industry, though in the end it is the economics that will be the deciding factor (Table 3). The most important issues identified:

- Assessment of previous patterns of market development in similar markets.
- Increased engagement of utilities and large energy companies.
- National energy plans and government support for renewable energy.
- Technical development.
- Growth in market and the present dynamics of the industry.
- Information about specific large projects.
- Assessment of wind resources and how they can be used.

Most of these factors are favourable for the industry at the moment. There is strong political support for wind energy, both as engineering and supply chain problems that have been associated with rapid growth in the past.

Economic projections are difficult at the best of times, when economies are relatively stable and a reference 'business as usual' case can be used. However, there are numerous signals that the world faces very turbulent economic conditions for a while - a credit crunch may make some project finance difficult and the shortage of raw materials could lead to supply chain difficulties. However, the rapidly escalating price of oil is focusing a lot of attention on the price of energy and the hedge of electricity supply without a fuel cost is likely to become increasingly attractive to many companies and utilities. At some stage, rising fuel costs could lead to demand for wind energy becoming almost infinite. The main factors expected to influence the continuing growth of the wind sector are:

- The economies of the transition states (Russia and Central Asia) will start to grow.
- Increasing energy demand in Asia and South America.
- Oil prices will continue to remain high as will demand for fossil fuels.
- Continuing competitiveness of wind with fossil fuels.
- Many countries may find they are well off their international CO_2 reduction commitments and need to install some new renewable capacity very quickly.
- Security of supply questions will continue to support wind power.
- Deregulated markets will remove excess conventional power capacity and new capacity is likely to be more expensive than wind.

Table 3. Market shares 2005-2007

Years		2005		2006		2007	
Manufacturer	Country	Supplied	Share	Supplied	Share	Supplied	Share
Vestas	Denmark	3186	27.6%	4239	28.2%	4503	22.8%
Ge Wind	US	2025	17.5%	2326	15.5%	3283	16.6%
Gamesa	Spain	1474	12.8%	2346	15.6%	3047	15.4%
Enercon	Germany	1640	14.2%	2316	15.4%	2769	14.0%
Suzton	India	700	6.1%	1157	7.7%	2082	10.5%
Siemens	Denmark	629	5.4%	1103	7.3%	1397	7.1%
Acciona	Spain	224	1.9%	426	2.8%	873	4.4%
Goldwind	China	132	1.1%	416	2.8%	830	4.2%
Nordex	Germany	298	2.6%	505	3.4%	676	3.4%
Sinovel	China	3	0.0%	75	0.5%	671	3.4%
Others		1032	8.9%	1094	7.3%	2076	10.5%
Total		11343	98%	16003	107%	22207	112%

While wind energy can still seem a small industry compared with conventional power generation, the achievement of 1% of world electricity generation is potentially significant. In individual markets such as Denmark, Germany and Spain reaching 1% has been a breakthrough figure, establishing a critical mass and being followed by further rapid growth in each year market. If the same pattern is seen with world wind energy demand and the industry continues to establish itself as a significant player in the energy sector and pushes on rapidly to 30% of world electricity demand and beyond, then the glass should be seen as half full.

Wind energy is one of the low investments high yielding sources of power generation. The future of wind energy is extremely bright and there is no doubt that in the renewable energy sector, wind power would play a predominant role in adding to the national grids clean and non-polluting energy in the coming years (Table 4).

In recent years, demand for the micro wind turbines, of the output below 1 kW, is on the increase as monuments and educational materials. Most of the micro wind turbine that has a diameter under 1.0 m is low blade tip speed ratio type on the market, by the problem of the frequency, the safety and the blade noise. In these circumstances, it would be necessary to develop the system characteristics of micro wind turbines for the purpose of much higher performance in spite of the low Reynolds number regions. Wind power generation is characterised by its stochastic nature, whereby supply and demand, in small grid systems in particular, mostly do not match. The combination of wind power with a second complementary power generation and/or direct/indirect storage technology therefore has, in principle, considerable potential. Wind-diesel, wind-water desalination and wind power in combination with hydrogen production are all potential options that have been high on the international renewable energy agenda for several years. A small-scale wind-PV hybrid power generator system for dairy farm is shown in Figure 5, to verify the possibilities to apply a power generating system and heating source for dairy farm. It is possible to apply the system for power supply and heat source to melt snow and process fertiliser.

Table 4. Installed capacity per year

Year	Europe (MW)	World (MW)
Before 2000	9.413	13.954
2000	13.306	18.449
2001	17.812	24.927
2002	23.832	32.037
2003	29.301	40.301
2004	34.725	47.912
2005	40.897	59.320
2006	48.628	74.517
2007	57.136	94.593
2008	66.785	120.458
2009	78.514	151.753
2010	93.590	191.318

4. SOLAR ENERGY

Global investment in renewables and energy efficiency now outpaces that for nuclear energy. Renewables also accounted for more than a fifth of new generation capacity built in 2007. The renewable energy and energy efficiency sectors seeing a level of commercial investment that most thought unattainable just a few years back. The risk factors pointing towards a 'bust' as identified and summed up in Figure 6.

Studies have been begun to estimate both the economic effects that climate change will have on global society as well as the costs of possible climate change mitigation and adaptation measures. Although the capacity to enact either a mitigation or adaptation strategy is based on country-specific conditions, technology and information availability, models have been used to calculate the approximate cost to stabilise atmospheric emissions at different levels.

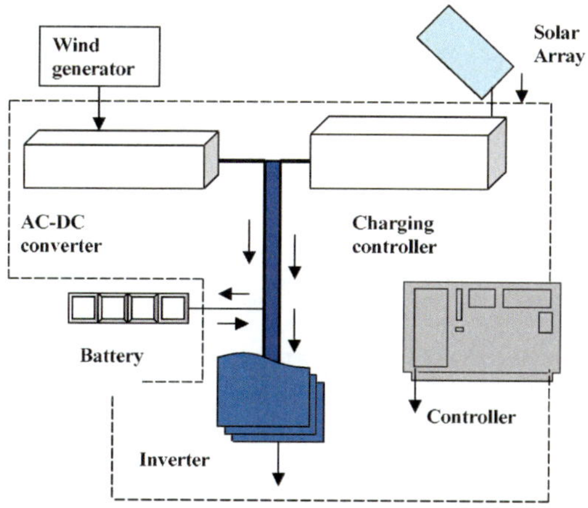

Figure 5. Wind-Photovoltaic hybrid generation systems

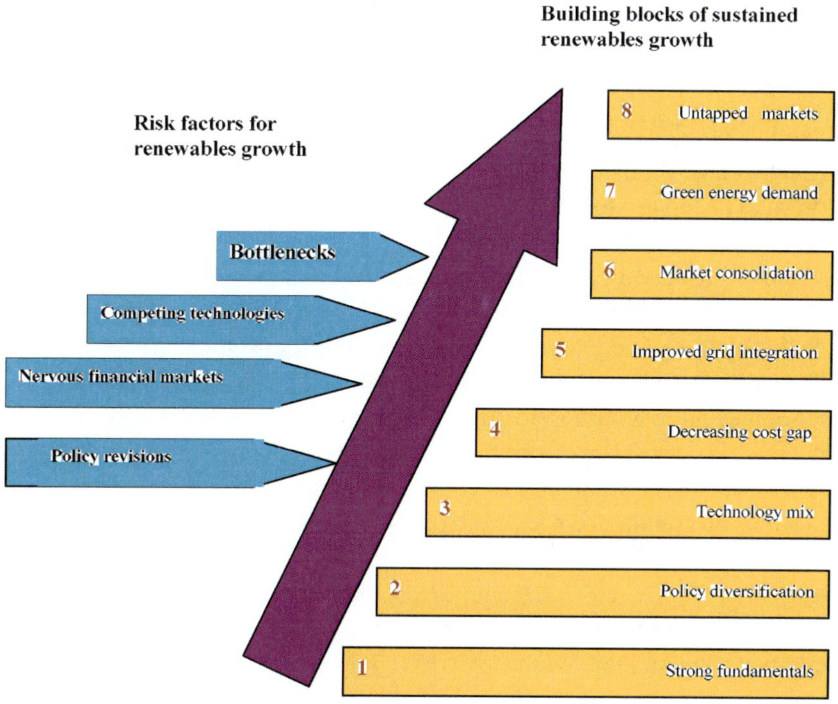

Figure 6. Summary of risk and risk mitigation factors

Wind power is far from the only clean energy sector on the rise and many of the technologies following in its tracks are much more decentralised, including roof-top systems like photovoltaics (PVs) or solar thermal and energy efficiency technologies on the demand side. Solar and energy efficiency were actually the two largest sectors in terms of venture capital investment, with solar bringing in 30% and efficiency 18% (Figure 7). Besides the high level of early stage investment, mostly focusted on new technology development, these two sectors also fared well on the public stock markets, ranking second and third after wind. Solar would have overtaken wind on the public markets.

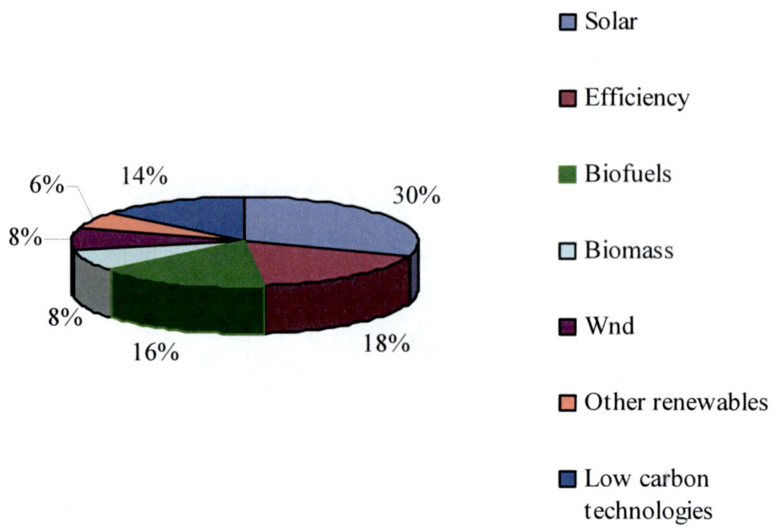

Figure 7. Investment in renewable and energy efficiency

The potential of electric power generation from incorporating PVs in buildings is enormous. If the electrical power demand of many countries is to be supplemented by the use of PV, it is deemed necessary to integrate such systems into many of the building faces. Many larger structures such as superstores, public buildings and most houses use mass produced tiles on their roofs. Such areas lend themselves useful in contributing to the energy used in the building or to export to the electrical grid when active roof tiles (PV-tiles) are introduced as a part of the roof structure. The integration of PVs within both domestic and commercial roof offers the largest potential market for PV especially in the developed world. Numerous national programmes are attempting to stimulate this market using standard PV modules as a roof element. However, roofs are not static, uniform structures and so are not ideally suited to modules which require precise, planar mounting structures. To date, attempts at producing a PV roof tile which accommodates current roof practice, in terms of both installation and aesthetics. The market for PV has historically been based on off-grid application where the relatively high cost of PV could be economically justified. In 2001 about 330 MWp of PV were produced and installed around the world and the growth rate of the industry is over 30% per year-100% per year in some countries with aggressive implementation schemes (Figure 8).

There have now been a number of successful large-scale programmes of systems deployed for basic power needs in rural households in developing and less developed countries. Remote applications servicing other applications such as telecommunications, cathodic protection, water pumping, etc., continue to grow as well. The PV market will continue to grow strongly for the next several years at least, driven by incentive programmes, cost reductions and greater market awareness.

It is becoming evident, that the quality will be the key to the PV market (Figure 9). The European Commission's Altener Programme a Training Manual was developed by the Global Approval Programme for PVs (PV GAP) to help manufactures of PV products to introduce quality management in their production. The manual contain important up-dates of the PV GAP Manuals for PV manufactures, published by the World Bank in 1999, including alignment with the 2000 edition of ISO 9001. This revised training manual was also translated from English into French, German and Spanish.

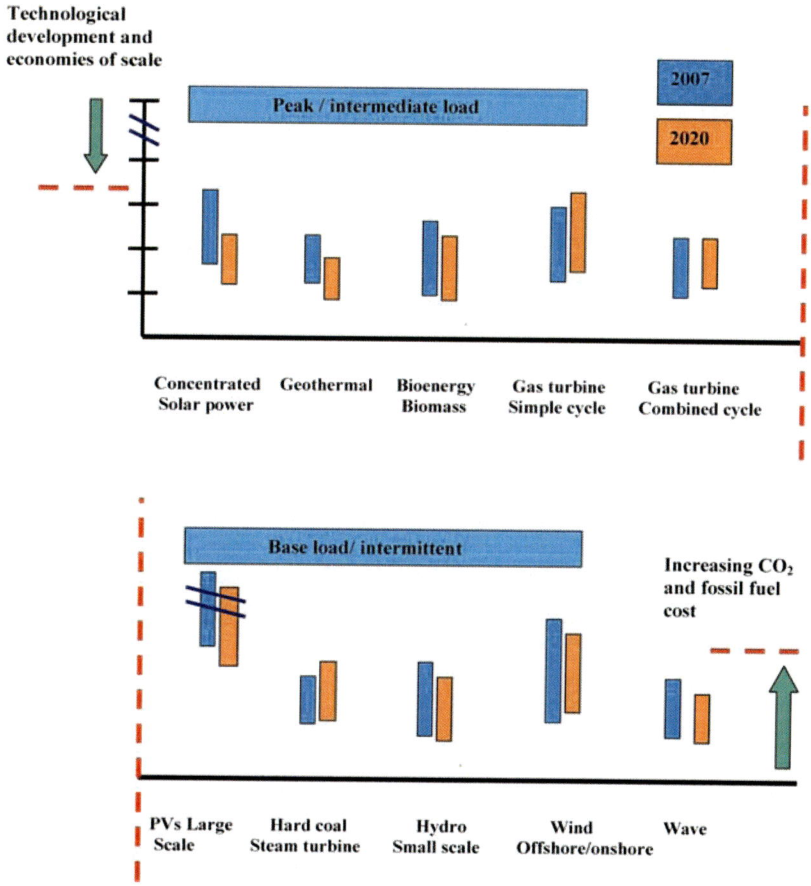

Figure 8. Cost increase for fossil primary energies and CO_2 emissions and continued technological development and economies of scale in renewables will improve competitiveness

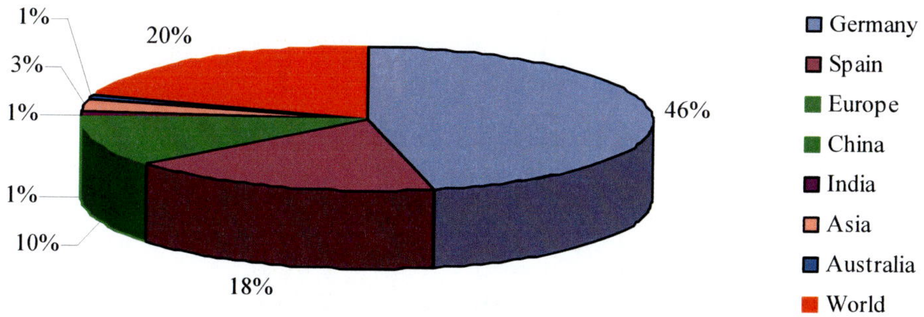

Figure 9. Differences in predicted PV market volumes worldwide until 2010

Water pumping is one of PV modalities that are growing in rural areas, mainly in developing countries. Due to the importance for health and food production it may be regarded as one of the noblest solar PV uses in isolated areas (Figures 10-11). Reliability and autonomy or self-reliance is understood as main factors for this high growth rates.

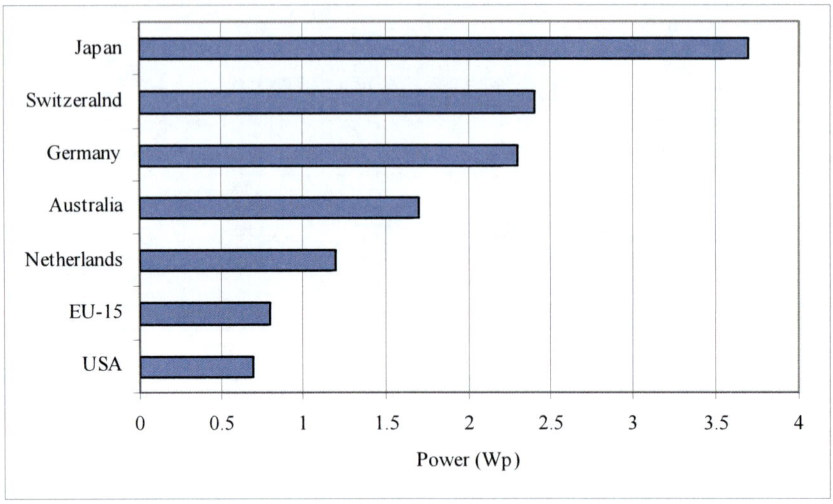

Figure 10. PV distributions for different countries per person

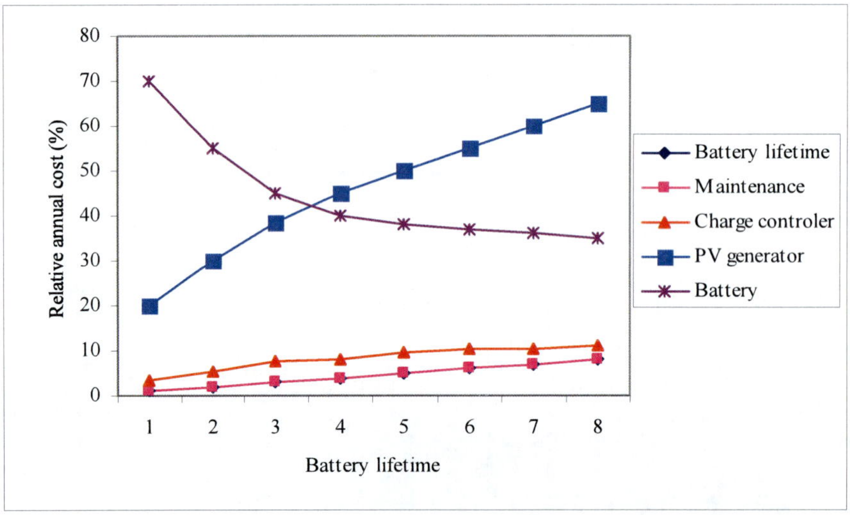

Figure 11. Relative annual costs of the components of small PV stand alone system with a lead-acid battery

5. Hydropower Potential

This section discusses various aspects of hydropower including: harnessing ocean energy, hydroelectric dams and micro hydropower systems. This section on hydropower explores the factors associated with utilising the actual potential of hydropower energy. The section covers all the technological details, along with issues and challenges faced during the utilisation of hydropower energy. Major projects, power plants, players in the industry, the major role of the United States in the global hydropower industry, and the various environmental benefits of using hydropower energy are all explored in depth in this section.

The growing worldwide demand for renewable energy projects is being driven by ever increasing global energy consumption and the availability of carbon and renewable energy credits. Renewable energy is entering a new phase with additional funding becoming available from governments, from socially responsible equity funds, and from public capital raisings.

Hydropower is the capture of the energy derived from moving water for some useful purpose. Prior to the widespread availability of commercial electric power, hydropower was used for irrigation, milling of grain, textile manufacture, and the operation of sawmills.

Hydropower produces essentially no carbon dioxide or other harmful emissions. In contrast to burning fossil fuels, this energy is not a significant contributor to global warming through production of CO_2.

Hydroelectric power can be far less expensive than the electricity generated from fossil fuel or nuclear energy. Areas with abundant hydroelectric power attract industry. Environmental concerns about the effects of reservoirs may prohibit development of economic hydropower sources in some areas.

Hydropower currently accounts for approximately 20% of the world's electricity production, with about 650,000 MW installed and approximately 135,000 MW under construction or in the final planning stages. Notwithstanding this effort, there are large untapped resources on all continents, particularly in areas of the world that are likely to experience the greatest growth in power demand over the next century. It is estimated that only about a quarter of the economically exploitable water resources has been developed to date, leaving the potential for hydro to continue to play a large role in sustaining renewable global electricity production in the future.

Apart from a few countries with abundance, hydro power is normally applied to peak load demand because it can be readily stopped and started. Nevertheless, hydroelectric power is probably not a major option for the future of energy production in the developed nations, however, because most major sites within these nations are either already being exploited or are unavailable for other reasons, such as environmental considerations.

Future hydropower energy programmes must be put into practice in conjunction with sound policies that restrict the use of fossil fuels and natural resources and contribute to the reduction of emissions into the environment. Such a strategy should be based in a sound scientific basis, without ideology, politics or financial interests. It should be implemented on a worldwide basis and not limited to industrialised countries. To achieve this goal, existing hydropower energy options must be evaluated for implementation, new strategies must be formulated and new, innovative solutions have to be found.

All projects are required to have environmental impacts assessment conducted, covering all potential damage to the environment, mitigation and restoration, a reclamation plan including a resettlement programme for displaced residents, and the estimated implementation costs. All hydro projects are required to conduct an environmental impact study. Water is essential to industry for processes such as cooling, cleaning, diluting and sanitation. With increasingly stringent water abstraction limits, recent droughts and a growing interest in the environmental performance of businesses, there is a need for industry to reduce water use.

6. GASIFICATION

An important consideration for operators of wastewater treatment plants (WWTPs) is how to handle the disposal of the residual sludge in a reliable, sustainable, legal and economical way. This by-product of wastewater treatment contains abundant organic material, including many kinds of bacteria. It also contains heavy metals and its composition is generally unknown. The benefits of drying sludge can be seen in two main treatment options:

- Use of the dewatered sludge as a fertiliser or in fertiliser blends.
- Incineration with energy recovery.

Use as a fertiliser option takes advantage of the high organic content 40-70% of the dewatered sludge and its high levels of phosphorous and other nutrients. However, there are a number of concerns about this route including:

- The chemical composition of the sludge (e.g., heavy metals, hormones and other pharmaceutical residues).
- Pathogen risk (e.g., salmonella, Escherichia coli, prionic proteins, etc.).
- Potential accumulation of heavy metals and other chemicals in the soil.

Sludge can be applied as a fertiliser in three forms:

- Liquid sludge.
- Wet cake blended into compost.
- Dried granules.

Use as energy recovery option takes advantage of the energy available in the sludge's organic content. Drying the sludge reduces its water content, thus increasing its calorific value and making it easier to combust. It also reduces odours and improves handling, with lower transport and storage costs. Sludge from WWTPs is typically combusted in (Table 5):

- Cement kilns.
- Coal-fired power plants.
- Mono-incinerators, i.e., plants burning refuse-derived fuel or a single waste stream.
- Mixed waste incinerators, e.g., municipal waste incinerator.

The recognised advantages of energy recovery from sludge include:

- The high calorific value (similar to lignite) of dewatered sludge.
- The use of dewatered sludge as a carbon dioxide (CO_2) neutral substitute for primary fuels such as oil, gas and coal.
- The use of dewatered sludge is a 'sink' for pollutants such as heavy metals, toxic organic compounds and pharmaceutical residues, thus offering a potential disposal route for these substances provided the combustion plant has adequate flue gas cleaning.

- The potential, under certain circumstances, to utilise the inorganic residue from sludge incineration (incinerator ash), such as in cement or gravel.

Table 5. Energy recovery from WWTP sludge

Form of sludge and usage	Cement kiln	Coal-fired power plant	Mono-incinerator	Refuse incinerator
Dewatered sludge	No	Yes	Yes/No*	Yes
Fuel substitute	-	No	No	No
Thermal dry biosolids	Yes	Yes	Yes	Yes
Substitute for fuel	Yes	Yes	Yes/No	No
Substitute for minerals	Yes	No	No	No
Type of process	Residue-free process	Fuel substitution	Disposal	Disposal

*Depending on type of plant

The demands placed on the drying system are therefore critical and include:

- High process stability.
- High mechanical reliability.
- High safety standards under all operating conditions.
- Compliance with environmental legislation such as emission limits.
- A product with properties suitable for a wide range of uses.

The energy efficiency formula takes into account the energy generated by the plant and puts it in relation to the calorific value of the municipal waste [2]. The energy introduced into the process from outside (such as fossil fuels or electricity) is subtracted. The energy efficiency can be improved by, for instance, reducing the input of fossil fuels. According to the European commission's formula, the energy efficiency for waste-to-energy (WTE) plants is calculated as follows:

$$\text{Energy efficiency} = E_p - (E_f + E_i) / 0.97\,(E_W + E_f) \qquad (1)$$

Where:

E_p is the annual energy produced as heat or electricity in GJ/year. It is calculated with energy in the form of electricity multiplied by 2.6 and heat produced for commercial use multiplied by 1.1.

E_f is the annual energy input to the system in GJ/year from fuels contributing to the production of steam.

E_W is the annual energy in GJ/year contained in the treated waste calculated using the lowest net calorific value of the waste.

E_i is the annual energy imported in GJ/year, excluding E_W and E_f. For thermodynamic reasons, E_f must be deleted in the nominator of the equation as it is included twice- in the nominator and the denominator.

An energy efficiency factor of 0.6, which has been proposed by commission, is too high for most existing plants. A threshold of 0.5 would be sufficient, with a further reduction of 0.1 for small plants and plants that produce electricity only due to a lack of demand for heat.

A factor of 0.6 would disadvantage smaller plants as they generally need the same energy for operation as larger plants, but have a lower throughput. It should be noted, at this point, that the public tend to prefer smaller plants in order to reduce the distance waste is transported (Figures 12-13).

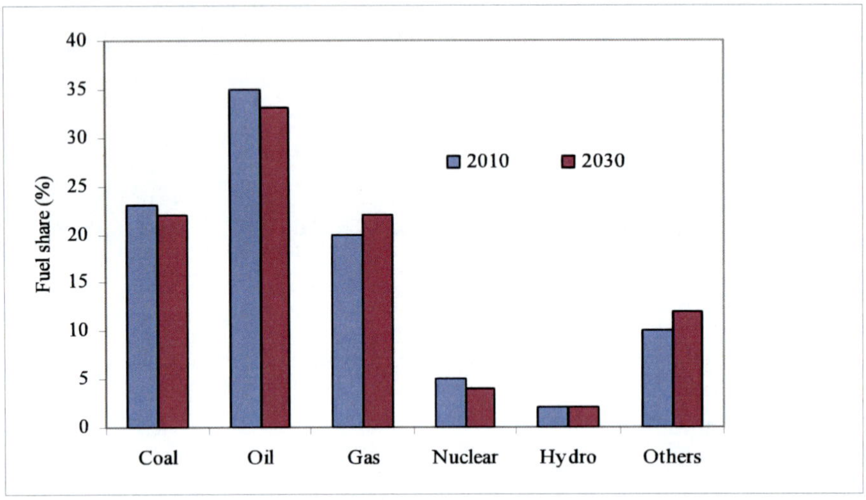

Figure 12. Outlook for world total primary energy supply

('Others' includes combustible renewables and waste, geothermal, wind and tidal energy)

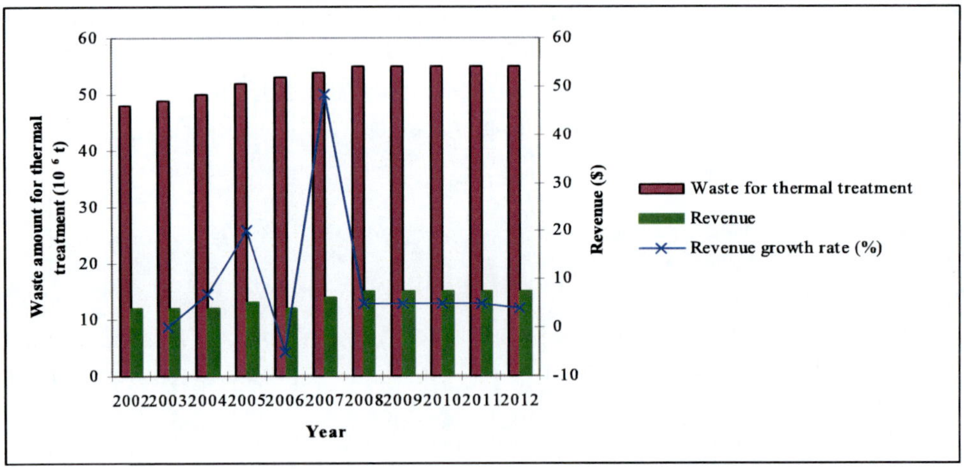

Figure 13. Volume shipments and revenue forecasts for the western European thermal waste treatment services

7. Energy Recovery

Mechanical-biological treatment (MBT) can be one option for improving the conservation of resources and energy in waste management systems. Mechanical-biological treatment enhances the conservation of embedded energy through recycling and allows potentially more efficient combustion or conversion of refuse-derived fuel (RDF). MBT encompasses a wide range of technologies aiming to process solid waste by a mixture of mechanical and biological separation. It also enables metals and other dry recyclables to be recovered (Figure 14). There are five main types of MBT process:

- Incorporating anaerobic digestion to generate biogas for electricity production. Anaerobic digestion also generates a digestate to be discharged or to be dewatered, producing a compost product.
- Producing an RDF product.
- Producing a compost product and/or a stabilised material for landfilling as well as a RDF product.
- Producing a compost product.
- Stabilising waste prior to landfill.

Gas production in the gasification system is controlled in two ways:

- The system analyses the gas produced and proprietary software feeds instructions back to a control system, delivering constant management of operating variables.
- A surge and mixing tank blends the gas flow.

Social and environmental will benefit to the community from the utilisation of alternative energy and reduced fossil fuel consumption.

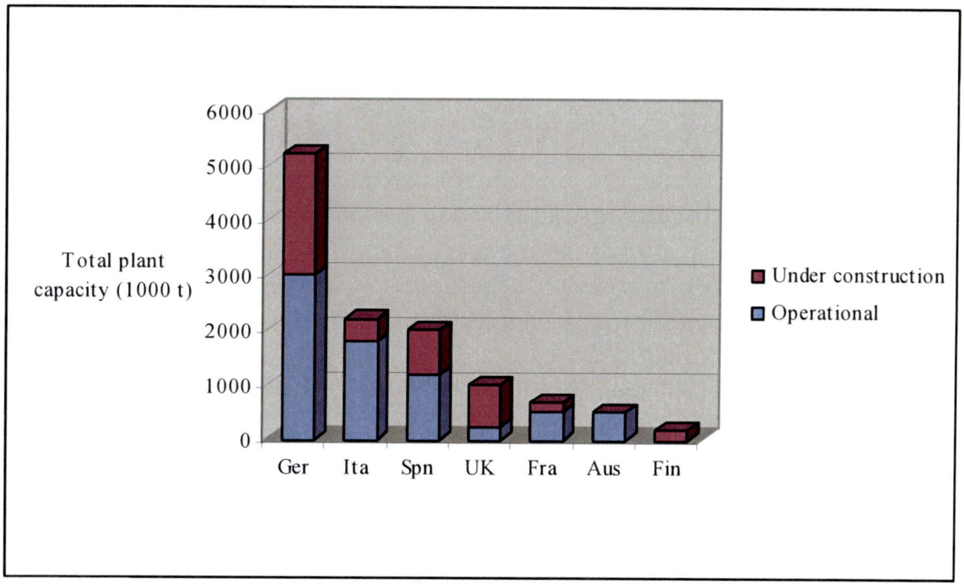

Figure 14. Capacity of plants in each country

Recycling is hugely beneficial, both from an economic and environmental perspective. The environment and economy will both benefit from an improved and more efficient reverse supply chain. This, in turn, would result in increased and optimised recycling of waste streams that reduce the depletion of scarce natural resources- water, oil and minerals- while at the same time protecting the environment by keeping these valuable materials in service. Figure 15 shows how the individual LCAs performed on avoided greenhouse gas emissions in CO_2 equivalent by material [3]. A cornerstone of recycling is effective waste collection and many local authorities are now looking to households and businesses to ensure this is in place. It is no longer sufficient to leave one or more bags of 'rubbish' at the end of a drive or back of a workplace.

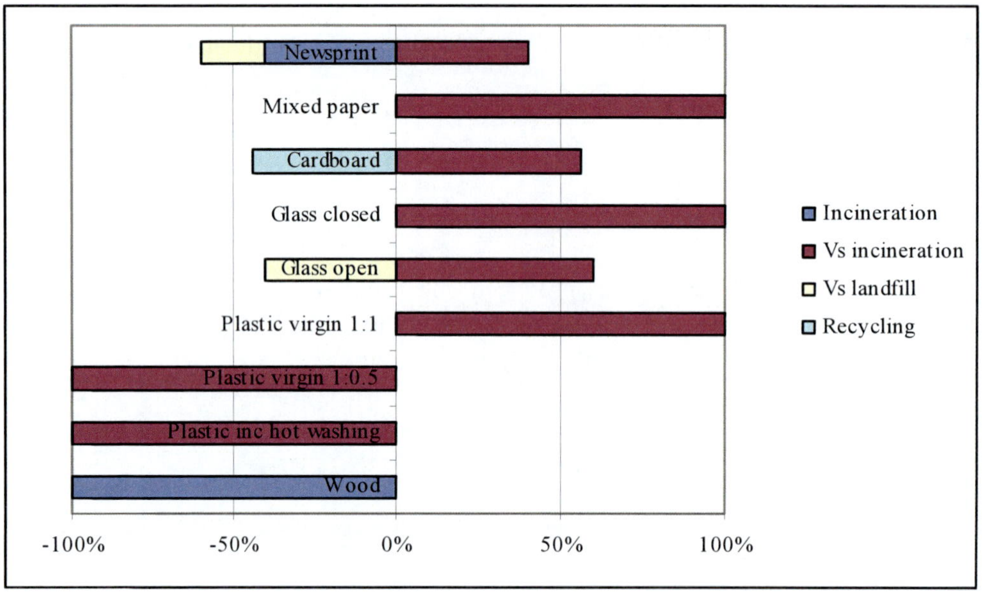

Figure 15. Performance of the life cycle analysis (LCA) on avoided greenhouse gas emissions

Odorous emissions are a serious concern relating to biowaste treatment facilities and affect the likelihood of planning permission being granted particularly in urban areas. Research has shown odour is one of the prime concerns from urban residents where MBT plants are proposed. An often unrecognised aspect of the waste management field is the many examples of individuals giving back. A conventional biofilter operates by engineering the correct environment in a closed vessel, ensuring that the bacteria within it are cultivated to effectively biodegrade the compound in the air that passes through it. Recirculation of the airstreams improves the rate of degradation by the bacteria. This has several advantageous characteristics:

- A surface s structure that is excellent for supporting bacteria and has the ability to retain moisture in the event of temporary water failure.
- A self-supporting structure with good packing characteristics that ensures there is minimal pressure drop across the media bed (thus reducing running costs).
- The right conditions to encourage bacteria to metabolise more than grow, which minimises sludge production and hence associated disposal.

8. Geothermal Heat

This study explores the factors associated with utilising the actual potential of geothermal energy. It also covers all the technological details, along with issues and challenges faced during the utilisation of geothermal energy. Major projects, power plants, players in the industry, the major role of the United States in the global geothermal industry, the active role of the US Department of Energy, and the various environmental benefits of using geothermal energy are all explored in-depth in this study.

Geothermal power is the use of geothermal heat to generate electricity. Geothermal comes from the Greek words geo, meaning earth, and therme, meaning heat. The utilisation of geothermal energy for the production of electricity dates back to the early part of the twentieth century. For 50 years the generation of electricity from geothermal energy was confined to Italy and interest in this technology was slow to spread elsewhere. In 1943 the use of geothermal hot water was pioneered in Iceland. Estimates of exploitable worldwide geothermal energy resources vary considerably.

The largest dry steam field in the world is the geysers, about 90 miles (145 km) north of San Francisco. The geysers began in 1960 which has 1360 MW of installed capacity and produces about 1000 MW net. Calpine Corporation now owns 19 of the 21 plants in the geysers and is currently the United States' largest producer of renewable geothermal energy. The other two plants are owned jointly by the Northern California Power Agency and Santa Clara Electric. Since the activities of one geothermal plant affects those nearby, the consolidation plant ownership at the geysers has been beneficial because the plants operate cooperatively instead of in their own short-term interest. The geysers are now recharged by injecting treated sewage effluent from the City of Santa Rosa and the Lake County sewage treatment plant. This sewage effluent used to be dumped into rivers and streams and is now piped to the geothermal field where it replenishes the steam produced for power generation.

Another major geothermal area is located in south central California, on the southeast side of the Salton Sea, near the cities of Niland and Calipatria, California. As of 2001, there were 15 geothermal plants producing electricity in the area. Cal Energy owns about half of them and the rest are owned by various companies. Combined the plants have a capacity of about 570 megawatts. Geothermal energy can be used as an efficient heat source in small end-use applications such as greenhouses, but the consumers have to be located close to the source of heat. Geothermal energy has a major environmental benefit because it offsets air pollution that would have been produced if fossil fuels were the energy source. Geothermal energy has a very minor impact on the soil - the few acres used look like a small light-industry building complex. Since the slightly cooler water is reinjected into the ground, there is only a minor impact, except if there is a natural geyser field closed by. The world's decentralised energy industry is right to look towards developing countries as potential markets for its wares. These countries have the opportunity to miss out the industrial development stage characterised by a single, centralised electricity transmission and distribution system, in favour of a hybrid of local grids liberally supplied by small-scale decentralised power generators.

9. Greenhouse Gas Emissions

The typical challenge in the developing world is lack of everything, including electricity. When striving to feed the increasing power demand to enable social reform and industrial growth, local decision makers face the question of which route to take. Shall we copy the model of the rich countries, or could there be another, maybe better way? The need to reduce CO_2 emissions presents a new, additional challenge, difficult even for the richest of nations.

Conventional, centralised electricity networks are the norm in the developed world. However, the present energy infrastructure of the developed countries was mainly created during the monopolistic utility era of the past. The utilities had the power to decide what kind of capacity to construct and how to construct the grid. There was practically no competition. The main challenge for many utilities was to get construction permits for building new generation capacity- the construction itself was practically risk free as they could turn their cost structure into a solid power tariff. Capital was easily available and cost competitiveness and overall system cost optimisation were not the main concerns.

Today the market situation and rules have permanently changed. Progressive modern utilities have left the past behind and are striving towards a modern, competitive energy system. However, one remainder of the past, still to some extent maintaining the economies of scale thinking, is to look at and calculate power plant and grid investments separately, as if they had nothing, or very little, to do with each other. Installing flexible, high-efficiency peaking and grid stability generation capacity closer to the load pockets of cities and industrial areas would be a natural way to go, but it does not fit the model of old, large-scale utility thinking (Figure 16). A typical electricity system built by a rich country monopoly utility 'wasted' capital in the following ways:

- Large, steam-fired power plants, running on part load, were used for frequency and load control (unless hydro non-dispatchable wind condition, but the problem has been solved with the very strong grid, which can transmit the wind power over long distances to remote countries.
- Efficient peaking capacity was not constructed; instead, the so-called peaking plants were typically based on largest possible simple cycle industrial gas turbines located in critical points in grid. This capacity functional mainly as an emergency reserve at grid nodes and was hardly ever used as it has such poor heat rate and relatively long starting time.
- Excessive base-load capacity was constructed. This was possible as there was an ensured return on asset investments.
- The grid has been sized to transmit the full peak power from large remote power plants to the consumption centres in cities.

Utilities in the developing world face a major challenge in developing their electricity systems. The parameters to optimise at the same time are:

- Reliability of supply.
- System flexibility and preparedness for load growth.
- Economical competitiveness, i.e., cost effectiveness, elimination of 'wastes'.

- Access to power for the whole nation.
- Ensured access to fuels and fuel flexibility.

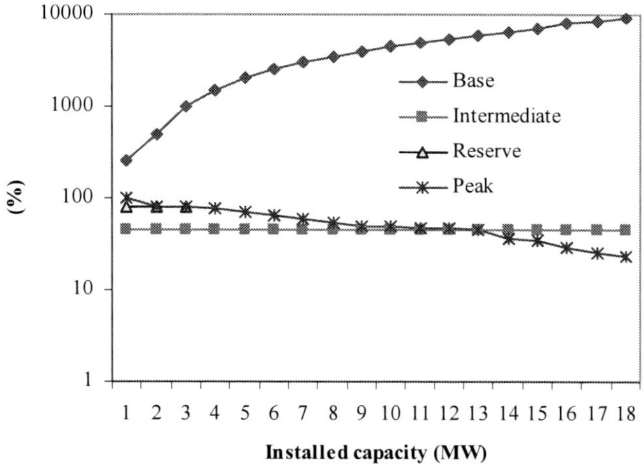

Figure 16. Load duration curve showing the base, intermediate, peaking load segments, and necessary contingency reserve

Combined heat and power (CHP) has been installed and used for many years as a highly efficient energy supply system to fulfil electricity and thermal energy needs in a range of applications around the world. As environmental issues, particularly those of climate change caused by greenhouse gases (GHGs), become as an effective means to reduce emissions is more pronounced than ever. However, despite CHP's significant contribution to the reduction of CO_2 emissions, calculation of its effects can result in preventing further dissemination of CHP. This is because, when an appropriate calculation method is used, there is a risk of under-estimating the effects of CHP and even, in some instances, estimating erroneously an increase in CO_2 emissions as a result of installing a CHP system. Proper estimation of the benefits of CHP will be vital in encouraging more businesses to install CHP systems.

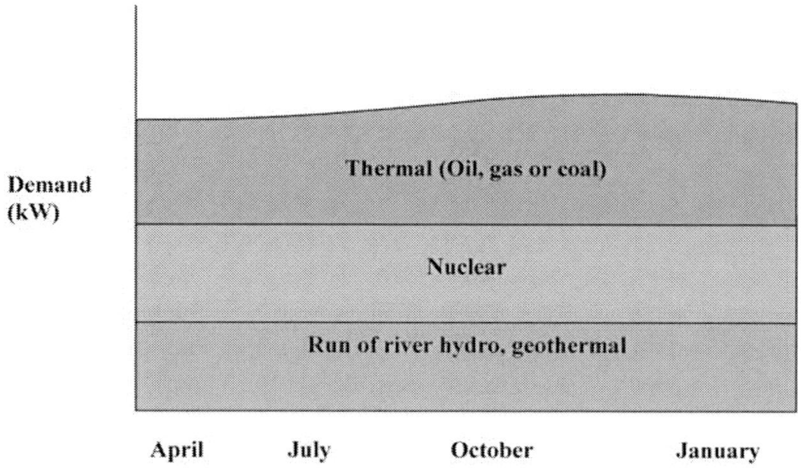

Figure 17. Image of annual load profile of power plants

Considerable debate surrounds the calculation of CO_2 emissions as a result of the use of the two emission factors [4-5]:

- All fuels average emission factor (AEF) - used to calculate emissions from power consumption.
- Marginal emission factor (MEF) - used to estimate reduced amount of CO_2 emissions as a result of lower power consumption.

CO_2 emission by consumption of a fuel is calculated as follows:

Emission amount (kg CO_2) = emission factor of a fuel (kg CO_2/MJ, kWh, etc.) x energy consumed (MJ, kWh, etc.) (2)

In the case of electricity, however, CO_2 is not emitted at the point of use. The power on the grid is usually generated by a mix of fuels – some CO_2 – emitting plants such as oil, coal and natural gas-based thermal power plants and non- CO_2 – emitting power sources such as nuclear, hydro and renewables. As a result, when calculating CO_2 emission by the use of grid power, the use of AEF or MEF is preferred. Therefore, CO_2 emission by an electricity user is estimated as follows:

CO_2 emission volume (kg CO_2) = AEF (kg CO_2/kWh) x power consumed (kWh) (3)

CO_2 emission of grid electricity is not accounted for on the demand side, rather on the generation side. AEF is obtained by dividing the total CO_2 emission volume from thermal power plants connected to the grid by the total power supplied through the grid including those that are free of emissions, indicating average CO_2 emission per kWh of power on grid.

AEF (kg CO_2/kWh) = total CO_2 emissions from all thermal plants on grid (kg/ CO_2)/total power supplied to grid from all sources (kWh) (4)

To calculate the amount of CO_2 emissions reduced because of lower power consumption, the emission factor for these plants must be multiplied by the reduced power consumption. The factor used is called the marginal emission factor or MEF. The reduced emission amount is obtained by the following calculation:

Emission reduction (kg CO_2) = MEF (kg CO_2/kWh) x electricity conserved (kWh) (5)

10. Energy Efficiency

Eventually renewable energies will dominate the world's energy supply system. There is no real alternative. Humankind cannot indefinitely continue to base its life on the consumption of finite energy resources. Today, the world's energy supply is largely based on fossil fuels and nuclear power. These sources of energy will not last forever and have proven to be contributors to our environmental problems. The environmental impacts of energy use are not new but they are increasingly well known; they range from deforestation to local and global pollution. In less than three centuries since the industrial revolution, humankind has already burned roughly half of the fossil fuels that accumulated under the earth's surface over hundreds of millions of years. Nuclear power is also based on a limited resource (uranium) and the use of nuclear power creates such incalculable risks that nuclear power plants cannot be insured.

Renewable sources of energy are an essential part of an overall strategy of sustainable development. They help reduce dependence of energy imports, thereby ensuring a sustainable supply. Furthermore renewable energy sources can help improve the competitiveness of industries over the long run and have a positive impact on regional development and employment. Renewable energy technologies are suitable for off-grid services, serving those in remote areas of the world without requiring expensive and complicated grid infrastructure. In 2007, the United States outlined plans to ease out of its foreign oil dependence through the use of renewable energy resources, and reduce gas usage by a full 20% in ten years through alternative fuels. Extending hope and opportunity depends on a stable supply of energy that keeps nation's economy running and environment clean.

Many wind/solar farms are located in remote areas served by low capacity radial distribution networks. These networks often have insufficient capacity to ship the power to demand centres when the turbines are generating at full capacity.

The public awareness of the depletion of the ozone layer has increased since 1970s. The ozone layer in the stratosphere protects life on earth against the ultraviolet radiation from the sun. Scientists and politicians argued for some years on the reasons for the reduced ozone layer, which was recognised especially over the Antarctic. The fact is that the ozone is depleted by the presence of chlorine. Further, very low temperatures as well as sunlight are required for having the process running. Vapour compression refrigeration systems using fluorocarbons, hydrocarbons or ammonia represent the established technology for household, commercial and industrial refrigeration and air-conditioning.

Understanding the earth and the processes that shape it is fundamental to the successful development and sustainable management of our planet. The evolution of the earth's crust over geological time has resulted in diverse, often beautiful, landscapes formed by earthquakes, oceans, fire and ice. These landscapes are also a source of mineral wealth and water, a base for engineering projects and a receptacle for the waste. Minerals are vital for manufacturing, construction, energy generation and agriculture. Some of the requirements can be met by increasing the use of recycled and renewable resources, but the need for new mineral resources continues. Minerals are important to maintaining the modern economy and lifestyle. Hence, everyone must make best use of these valuable assets whilst minimising the impact of their extraction on the environment. Global energy use will rise dramatically but world oil and gas production will eventually decline. Geological hazards account for huge loss of life and damage to property. Poorer countries often suffer the most due to inappropriate land-use planning and in the future climate change may worsen these effects. Clean drinking water is a basic human need. The environmental sustainability of water resources, especially when balancing ecological and human needs with economic growth, is of major concern to governments worldwide. Increases in the world's population, the growth of cities, industrial development and ever-increasing waste and pollution can increase pressure on the environment. Understanding the effects of climate change is a key issue for society and researches focus on how to predict future climate change events and how to mitigate them.

Two essential components for economic development are a soundly based knowledge of a country's natural resources, such as groundwater, minerals and energy, and an understanding of the geological environment. The latter includes the potential effects of earthquakes, volcanic eruptions, landslides and the forces of erosion and deposition on rivers and coastlines. Helping developing countries acquire this knowledge and apply it to promote economic growth, sustainable livelihoods and the protection of people. Sustainable land use involves protection of the natural environment (viable agriculture, forestry, water resources, and soil functions), the quality and character of the countryside and existing communities. Groundwater is an essential source of drinking water (70%). Although commonly regarded as pure, it is often vulnerable to pollution or may contain natural concentrations of elements that can adversely affect health. Population growth, intensive agriculture, urbanisation and higher standards of living have resulted in increasing, often unsustainable, demands on groundwater resources. These pressures and the likely effects of climate change mean that groundwater systems must be better understood so that they can be evaluated effectively, managed sustainably and protected securely.

District heating plays a very important role in transition economies. District heating systems in transition economies often face financial, technical or managerial problems largely created by an inadequate policy framework. These challenges include lack of customer focus, low efficiency, excess capacity, corruption and an uneven playing field. Figure 18 shows how these challenges are interrelated and create a vicious circle that undermines finances and competitiveness of district heating companies, jeopardising their long-term sustainability.

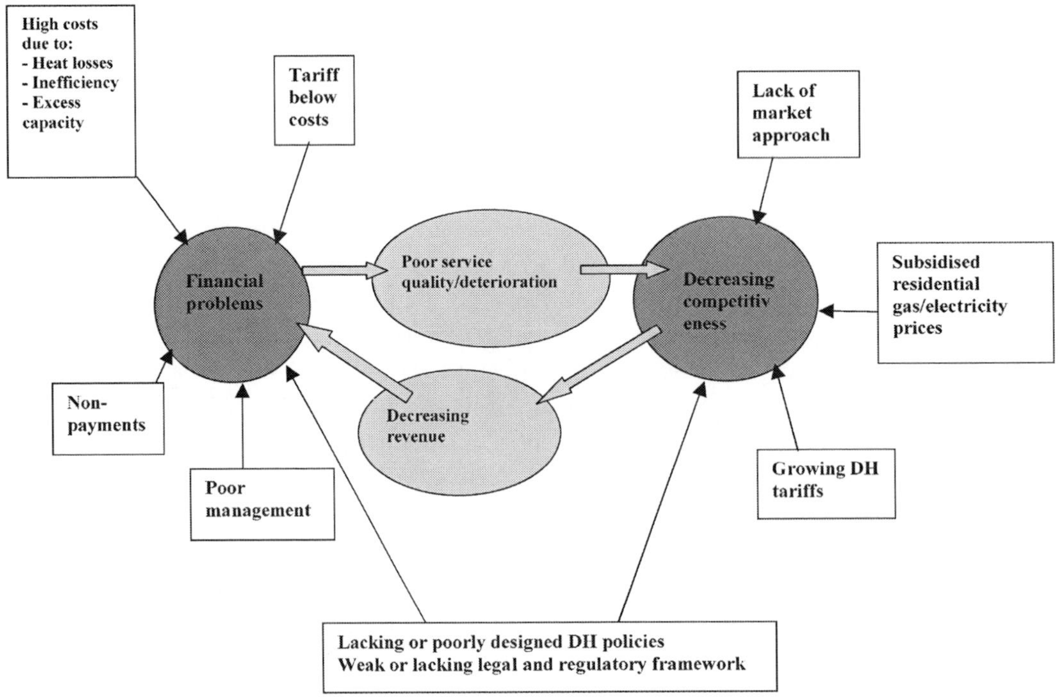

Figure 18. Key challenges for district heating systems in transition economies

10.1. Energy Supply

The earth is believed to be close to a state of thermal equilibrium where the energy, which is received at the surface by solar radiation, is lost again at night and the much smaller amount of energy, which is generated by the decay of unstable isotopes of Uranium, Thorium and Potassium distributed within the earth is balanced by the small continuous heat flux from the earth's interior to the pecans and atmosphere.

The use of geothermal energy involves the extraction of heat from rocks in the outer part of the earth. It is relatively unusual for the rocks to be sufficiently hot at shallow depth for this to be economically attractive. Virtually all the areas of present geothermal interest are concentrated along the margins of the major tectonic plates, which form the surface of the earth. Heat is conventionally extracted by the forced or natural circulation of water through permeable hot rock.

There are various practical difficulties and disadvantages associated with the use of geothermal power:

Transmission: geothermal power has to be used where it is found. In Iceland it has proved feasible to pipe hot water 20 km in insulated pipes but much shorter distances are preferred.

Environmental problems: these are somewhat variable and are usually not great. Perhaps the most serious is the disposal of warm high salinity water where it cannot be reinjected or purified. Dry steam plants tend to be very noisy and there is releases of small amounts of methane, hydrogen, nitrogen, amonia and hydrogen sulphide and of these the latter presents the main problem.

The geothermal fluid is often highly chemically corrosive or physically abrasive as the result of the entrained solid matter it carries. This may entail special plant design problems and unusually short operational lives for both the holes and the installations they serve.

Because the useful rate of heat extraction from a geothermal field is in nearly all cases much higher than the rate of conduction into the field from the underlying rocks, the mean temperatures of the field is likely to fall during exploitation. In some low rainfall areas there may also be a problem of fluid depletion. Ideally, as much as possible of the geothermal fluid should be reinjected into the field. However, this may involve the heavy capital costs of large condensation installations. Occasionally, the salinity of the fluid available for reinjection may be so high (as a result of concentration by boiling) that is unsuitable for reinjection into ground. Ocasionally, the impurities can be precipitated and used but this has not generally proved commercially attractive.

10.2. Refrigeration

The refrigeration industry has had to face environmental challenges. Regulation and behaviour related to the ozone layer differ from one country to another. Problems related to global warming are likely to cause similar problems in very different ways. These two kinds of problems may lead company managers to change/modernise equipment. These environmental factors must not hide other important development factors such as those related to hygiene, organoleptic quality, control methods and equipment packaging, etc. All equipment change and investment projects must take all of them into consideration. The more developed a country, is the more widely refrigeration is used. Food preservation is still the main use of refrigeration, followed by air conditioning, energy savings and transport (heat pump and liquefied gases), industrial processes and medicine. When consumers demand more and more fresh products, handling such products becomes increasingly difficult: cold stores, display cabinets, refrigerated trucks have to be much more effective. Difficulties are mostly related to:

- Very narrow ranges, in many cases, between temperatures involving microbial or chemical risk and temperatures that cause chilling injury or freezing.
- No or little overlapping between temperature ranges permitted for different products.
- No or little possibility for the consumer or the seller to evaluate the remaining life spans of the product or even to perceive any risk.
- Humidity control, vapour pressure differences have serious consequences, not only on the weight but also on the unit value of products.

The production and use of Chloro Fluoro Carbon (CFC) refrigerants have been phased out in developed countries. Hydro Chloro Fluoro Carbon (HCFC) refrigerants are now subject to regulation and are scheduled to be phased out in the twenty-first century. Recent developments in refrigeration technology and environment (climate) control have provided a better quality of life for mankind. As is known, refrigeration technology has played a key role in preserving the quality of food and eliminating waste (spoilage). The refrigerated transport of perishable products provides a critical link in the cold chain between the consumer and the producers, processors, shippers, distributors and retailers. The transport refrigeration industry incorporated mechanical vapour compression mechanisms into the first transport refrigeration units. Therefore, the physical, chemical and thermodynamic properties of these refrigerants had a major impact on the design of compressors and other refrigeration system components. Many refrigerants have excellent thermodynamic properties suitable for use in vapour compression refrigeration systems, but are limited by other properties such as toxicity, flammability and chemical stability within the refrigeration system.

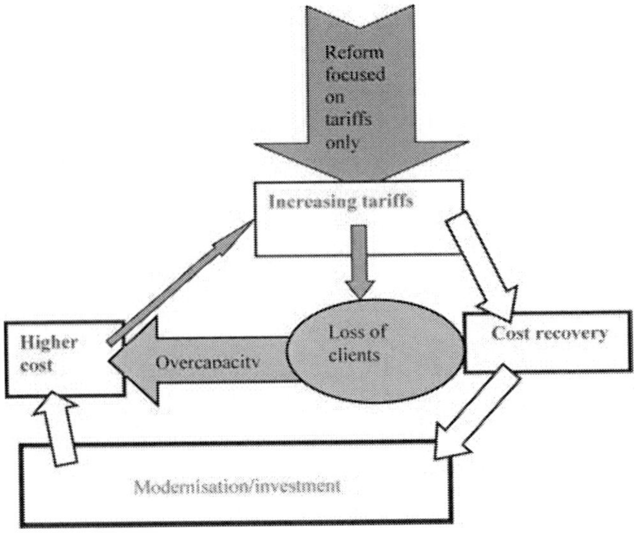

Figure 19. Unsustainable tariff growth

Most district heating systems in former planned economies are less efficient and oversized. In other words, their supply infrastructure is larger than necessary to meet current demand. This problem can be exacerbated when they lose customers. Reforms, particularly for tariffs, are needed to improve finances and break the vicious circle of deteriorating competitiveness. Yet, policymakers need to design the reforms very carefully for them not to backfire and make matters worse, as illustrates in Figure 19.

10.3. Temperature Distributions

World capacity of geothermal energy is growing at a rate of 2.5% per year from a 2005 level of 28.3 GW [1]. The GSHPs account for approximately 54% of this capacity almost all of it in North America and Europe [6]. The involvement of the UK is minimal with less than 0.04% of world capacity and yet is committed to substantial reduction in carbon emission beyond the 12.5% Kyoto obligation to be achieved by 2012. The GSHPs offer a significant potential for carbon reduction and it is therefore expected that the market for these systems will rise sharply in the UK in the immediate years ahead given to low capacity base at present. There are numerous ways of harnessing low-grade heat from the ground for use as a heat pump source or air conditioning sink. For small applications (residences and small commercial buildings) horizontal ground loop heat exchangers buried typically at between 1 m and 1.8 m below the surface can be used provided that a significant availability of land surrounding the building can be exploited which tends to limit these applications to rural settings. Horizontal ground loop heat exchangers can be used to circulate refrigerant (direct heat exchanger) or a water/antifreeze mixture (indirect heat exchange) and rely to some extent on solar input during summer for earth temperature recovery following winter heat extraction (Figures 20-21). For a more economical use of available land, a vertical ground loop heat exchange array or ''borehole'' array can be used typically involving a matrix of vertical borehole heat exchangers spaced at 5 m at depths of up to 180 m. high density plastic (typically high density polyethylene) tube of 20-40 mm nominal diameter is fed down to 100-150 mm diameter borehole to form a U-tube with the borehole subsequently grouted using a high conductivity hard-setting compound (usually bentonite). These and others methods of low grade-ground heat harvesting have been reviewed in details from the perspective of UK application by Rawlings [7].

Heat generation within the earth is approximately 2700 GW, roughly an order of magnitude greater than the energy associated with the tides but about four orders less than that received by the earth from the sun [8].

Temperature distributions within the earth depend on:

- The abundance and distribution of heat producing elements within the earth.
- The mean surface temperature (which is controlled by the ocean/atmosphere system).
- The thermal properties of the earth's interior and their lateral and radial variation.
- Any movements of fluid or solid rock materials occurring at rates of more than a few millimetres per year.

Figure 20. A photograph of a laying ground-loop heat exchanger

Figure 21. A photograph shows the earth loop vapour and liquid

Of these four factors, the first two are of less importance from the point of view of geothermal energy. Mean surface temperatures range between 0-30°C and this variation has a small effect on the useable enthalpy of any flows of hot water. Although radiogenic heat production in rocks may vary by three orders of magnitude, there is much less variation from place to place in the integrated heat production with depth. The latter factors, however, are of great importance and show a wide range of variation. Their importance is clear from the relationship:

$$\beta = q/k \tag{6}$$

Where:

β is the thermal gradient for a steady state (°C/km), q is the heat flux (10^{-6} cal cm^{-2} sec^{-1}) and k is the thermal conductivity (cal cm^{-1} sec^{-1} $°C^{-1}$).

The first requirement of any potential geothermal source region is that β being large, i.e., that high rock temperatures occur at shallow depth. Beta will be large if either q is large or k is small or both. By comparison with most everyday materials, rocks are poor conductors of heat and values of conductivity may vary from 2 x 10^{-3} to 10^{-2} cal cm^{-1} sec^{-1} $°C^{-1}$. The mean surface heat flux from the earth is about 1.5 heat flow units (1 HFU = 10^{-6} cal cm^{-2} sec^{-1}) [8]. Rocks are also very slow respond to any temperature change to which they are exposed, i.e., they have a low thermal diffusivity:

$$K = k/\rho C_p \tag{7}$$

Where:

K is thermal diffusivity; ρ and C_p are density and specific heat respectively.

These values are simple intended to give a general idea of the normal range of geothermal parameters (Table 6). In volcanic regions, in particular, both q and β can vary considerably and the upper values given are somewhat nominal.

Table 6. Values of geothermal parameters

Parameter	Lower	Average	Upper
q (HFU)	0.8	1.5	3.0 (non volcanic) ≈100 (volcanic)
k =cal cm^{-2} sec^{-1} $°C^{-1}$	$2x10^{-3}$	$6x10^{-3}$	$12x10^{-3}$
β =°C/km	8	20	60 (non volcanic) ≈300 (volcanic)

In many geothermal areas natural heat exchangers exist in the form of large-scale, sub-surface water circulation systems, which commonly give rise to hot springs or geyser activity at the surface (Table 7).

10.4. Thermodynamic Analysis of Refrigeration Cycles

Thermodynamics is the study of energy, its transformations and its relation to states of matter. A thermodynamic system is a region in space or a quantity of matter bounded by a closed surface. The surroundings include everything external to the system and the system is separated from the surroundings by the system boundaries. These boundaries can be movable or fixed, real or imaginary.

Table 7. Selected comparative cost data for geothermal energy [8]

Geothermal field	Geothermal production	Local average, other fuel
Electricity, US mills/kWh		
Iceland	2.5-3.5	-
Italy	4.8-6.0	~7.5
Japan	4.6	~6.0
Mexico	4.1-4.9	~8.0
Russia	7.2	~10.0
USA	5.0	7.0
Space heating, US$/Gcal energy		
Iceland	4.0	6.7
Hungary	3.0	11.0
Refrigeration, US$/Gcal		
New Zealand	0.12	2.40
Drying diatomite, US$/ton		
Iceland	~2	~12

Potential energy (PE) is caused by attractive forces existing between molecules, or the elevation of system.

$$PE = mgz \tag{8}$$

Where

m is the mass

g is local acceleration of gravity

z is the elevation above horizontal reference plane

[Net amount of energy added to system] = [Net increase of stored energy in system] (9)

Or

[Energy in] – [Energy out] = [Increase of stored energy in system] (10)

Refrigeration cycles transfer thermal energy from a region of low temperature T_R to one of higher temperature. Usually the higher temperature heat sink is the ambient air or cooling water, at temperature T_o, the temperature of the surroundings.

Performance of a refrigeration cycle is usually described by a coefficient of performance (COP), defined as the benefit of the cycle (amount of heat removed) divided by the required energy input to operate the cycle:

COP = useful refrigerating effect / Net energy supplied from external sources (11)

For a mechanical vapour compression system, the net energy supplied is usually in the form of work, mechanical or electrical and may include work to the compressor and fans or pumps. Thus,

$$COP = Qevap / Wnet \tag{12}$$

11. Environmental Challenge

The Montreal Protocol and its amendments (London 1990, Copenhagen 1992 and Vienna 1994) have given rise to the drawing up of regulation on the phase out the substances named Ozone Depleting Substances (ODS), including CFCs such as R12, R11 and HCFCs such as R22. Alternatives to CFCs include:

- Hydro Fluoro Carbons (HFCs) such as R134a and many mixtures.
- Naturally occurring refrigerants such as NH_3, hydrocarbons, water, air CO_2. Some of them are flammable and/or toxic or should be limited to relatively small niches.
- For other technologies such as absorption, solid sorption, etc., there are also some very interesting niches. However, refrigeration as a whole is not likely to shift there very soon because of the many other challenges it has to face.

The United Nations Framework Convention on Climate Change came into effect on March 21, 1994 and the Kyoto Protocol on December 10, 1997. Global warming is mostly caused by the rising percentage of various substances in the atmosphere, particularly CO_2, CH_4 and many others including CFCs, Hydro Chloro Fluoro Carbons (HCFCs) and HFCs. Refrigeration is involved because of:

- Refrigerants themselves, if they are released into the atmosphere, this is the 'direct effect'.
- CO_2 produced by energy consumption of the system throughout its lifespan, this is the 'indirect effect'.

The Fifth Framework Programme (FP5) is designed to ensure that European research efforts are translated more effectively into practical and visible results. The programme includes four thematic and three horizontal programmes.

Thematic programmes are:

- Quality of life and management of living resources.
- Users-friendly information society.
- Competitive and sustainable growth.
- Energy, environment and sustainable development.

Horizontal programmes are:

- Confirming the international role of community research.
- Promotion of innovation and encouragement of participation.
- Improving human research potential and the socio-economic knowledge base.

Research in agro-food industry, including refrigeration, has an important role, particularly under the key action ''Food, Nutrition and Health''.

11.1. Natural Disasters

The following is a detailed guide for communities and emergency operations team to develop and maintain a viable disaster management and recovery plan. In addition, it explains in details the concept of disaster management and various steps from planning to prepare against any disaster. It further highlights the role of government agencies and local authorities at the time of disaster as well as before and after it. It also discusses long and short-term goals for mitigation, planning and recovery from disaster along with initiatives that the US government has taken in recent years. It also, has a special focus on management of energy infrastructure during the time of disaster and presents a checklist for emergency response and recovery (Figure 22).

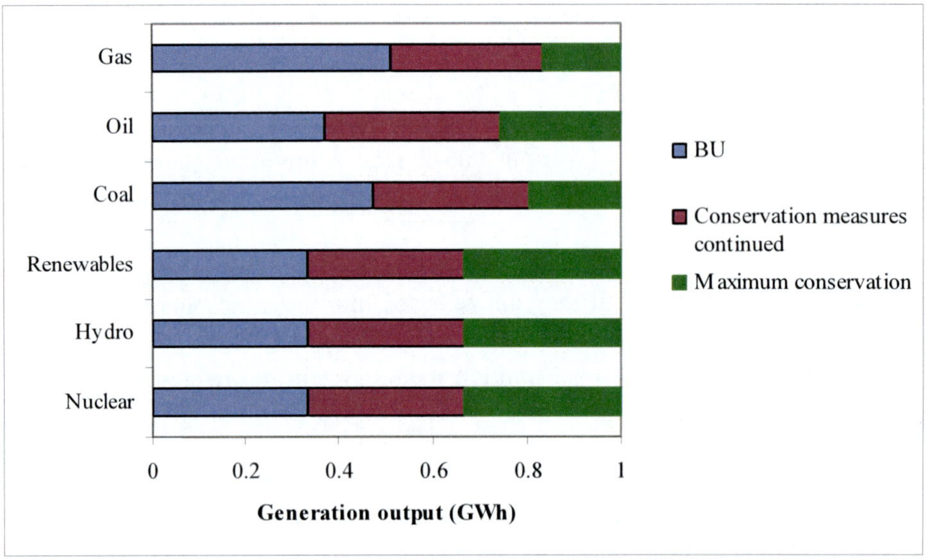

Figure 22. Energy conservation measures

Every year, tornadoes, hurricanes and other natural disasters injure and kill thousands of people and damage billions of dollars worth of property in the United States. Most of the times, it is almost impossible to prevent the occurrence of these disasters and their damages. However, it is possible to reduce their impact by adopting suitable disaster management strategy. Disaster management is a systematic approach towards preparing for disaster before it happens and includes disaster response - emergency evacuation, quarantine, mass decontamination - as well as supporting and rebuilding society after natural disasters have occurred. Efficient disaster management relies on thorough integration of emergency plans at all levels of government and non-government involvement.

Disaster preparedness, emergency management and post disaster recovery is highly dependent on economic and social conditions local to the disaster. However, the basics steps for disaster management remain same in all scenarios. Preparedness is the first step to counter disaster, which involves developing plan of action and it, includes communication, chain of command development, proper maintenance and training of emergency services and development of emergency warning systems along with emergency shelters and evacuation plans. Next step is response, which includes mobilisation of the necessary emergency services such as fire fighters, police, and ambulance that may be supported by a number of secondary emergency services, such as specialist rescue teams. Recovery from disaster involves restoration of the affected area including destroyed property, re-employment and redevelopment of essential infrastructure. Mitigation efforts attempt at preventing hazards from developing into disasters or reducing the impact of disasters and it focuses on long-term measures for reducing or eliminating future risks.

12. Conclusions

The following are concluded:

- Promoting innovation and efficient use of applicable renewable energy technologies.
- Identifying the most feasible and cost effective applications of renewable energy resources suitable for use.
- Highlighting the local, regional and global environmental benefits of renewable energy applications.
- Ensuring the renewable energy takes its proper place in the sustainable developments, supply and use of energy for greatest benefit of all, taking due account of research requirements, energy efficiency, conservation and cost criteria.
- Ensuring the financing of and institutional support for economic renewable energy projects.
- Encouraging education, research and training in renewable energy technology in the region.

12.1. Utilisation of Renewable Energy

Developing and implementing the use of renewable energy sources.

12.1.1. Policy and environment

- Efficiency, conservation, and policies.
- Renewable energy availability.
- Local and environmental concerns.
- Technology transfer.
- Financing requirements.
- Educational initiatives, legislative benchmarks.

12.1.2. Solar electrical technology

- PV technology manufacture, testing and certification.
- PV Stand-alone systems and components.

- PV for utility rural development and grid connection.
- PV Markets and commercialisation, financing schemes and national programmes.

12.1.3. Solar thermal technology

- Solar radiation prediction and analysis.
- Solar cooling, heating and rural applications.
- Solar thermal applications for power generation.
- Collector technology developments.

12.1.4. Solar and low energy architecture

- External Environment.
- Building, landscape design and urban communities and comfort productivity and health issues.
- Sustainable policy and social issues.
- Building simulation, material and design.
- Building refurbishment and integration of renewable energy; case studies.
- Internal environment.
- Thermal, ventilation and air movement.
- Lighting operation and control.

12.1.5. Wind energy technology and applications

- Small, micro-generation and hybrid systems.
- Machines and wind farms.
- Offshore wind power.
- Wind resources and environmental issues.
- Connection and integration.
- National and regional programmes.
- Economic and institutional issues.

12.1.6. Biomass conversion

- Heat and electricity generation.
- Energy crops and residues.
- Liquid fuels.
- Socio-economics, case studies, and environmental impacts.
- Gasification processes.

12.1.7. Fuel cells and hydrogen technology

- Fuel cells technology advances electricity generation.
- Fuel cells technology advances for transportation.

- Hydrogen production for fuel processing and systems.
- Proven commercialisation reports including economic and policy issues.

12.1.8. Marine/ocean energy

- Wave and tidal energy resources and their characterisation.
- Device modelling, testing and development.
- Device hydrodynamics, structural integrity and environmental analysis.
- Environmental impact assessment and standards.
- Balance of system - power take-off, sensors, controls and grid integration.
- Legislation, policy, finance and markets.
- Socio-economic assessment, education and training.

REFERENCES

[1] Brain, G. & Mark, S. (2007). Garbage in, energy out: landfill gas opportunities for CHP projects. *Cogeneration and On-Site Power, 8 (5),* 37-45.

[2] Federal Ministry for the Environment. Nature conservation and nuclear safety. Waste incineration- a potential danger. September 2005.

[3] WRAP. Environmental benefits of recycling: an international review of life cycle comparisons for key materials in the UK recycling sector. 2006

[4] Kyoto Protocol (CDM). http://cdm.unfccc.int/methodologies/Tools/EB35_repan12_ Tool_grid_emission.pdf

[5] GHG Protocol. http://www.ghgprotocol.org/files/electricity_final.pdf

[6] Lund, J. W., Freeston, D. H., and Boyd, T. L. (2005). Direct application of geothermal energy: 2005 Worldwide Review. *Geothermics*, 34, 691-727.

[7] Rawlings, R. H. D. (1999). Technical Note TN 18/99 – Ground Source Heat Pumps: A Technology Review. Bracknell. *The Building Services Research and Information Association.*

[8] Oxburgh, E. R. (1975). Geothermal energy. *Aspects of Energy Conversion*, p. 385-403.

In: Energy Costs, International Developments...
Ed: Leszek Kowalczyk and Jakub Piotrowski
ISBN: 978-1-60741-700-2

Chapter 3

BRIDGING RENEWABLE ENERGY AND RURAL COMMUNITIES: SOCIAL LEARNING AS A CATALYST

Lei Bi
School of Planning, Faculty of Environment, University of Waterloo,
Waterloo, Ontario, N2L 3G1, Canada

ABSTRACT

Social learning is a prerequisite factor of technical innovation and diffusion, particularly in cases where personnel and financial support are limited like developing rural areas. This article sets out some considerations from social learning theory that offers potentials of stimulating renewable energy technology diffusion through active public participation in rural areas. Findings from the experiences of eco-village construction in China's Hainan province are used to illustrate how a governmental top-down initiative is reinforced by a "*farmer-to-farmer extension*" process in promoting the adoption of renewable energy technologies or equipments, such as anaerobic digesters and solar water heaters. By exploring who learns, what is learned, why it is learned, how it is learned and impacts of social learning occurring in the technology diffusion process of an eco-village in Hainan, this paper confirms that farmers' cognition of future reinforcements in environmental soundness and economic savings can lead to active social learning and adoption of renewable energy technologies. Financial support from the government can catalyze the growth of the awareness. Moving towards more positive social learning in technology diffusion in developing rural areas, emerging lessons from rural Hainan are combined with propositions concerning adaptive planning, public participation in community development projects, skills and relations of village leaders, technical specialists, accessibility to information and sources of technologies and stakeholder collaboration.

Keywords: China, renewable energy, rural community development, social learning, public participation

INTRODUCTION

Energy is essential to economic and social development as well as improved quality of life, which are the fundamental aspirations of seventy percent of the people who live in the poor countries of Africa, Latin America, the Middle East, and Asia (Goldemberg, 1995; Dincer, 2000). Nowadays, poverty alleviation and improved well-being are two common and overarching development goals in most developing countries (Posa, Diesmos, Sodhi, & Brooks, 2008). Ironically, the vicious circle of poverty and environmental degradation is often mentioned in the "*sustainable development*" discourse in the Third World where population increase and poverty push people to degrade local resources and further impoverish themselves (Mink, 1993; Pearce & Warford, 1993). The implication of the focus on the vicious circle of poverty and environmental degradation in developing rural areas is not only the dialectic assumption of poverty alleviation will necessarily reduce environmental degradation, or its inverse, that arresting and reversing environmental decline will help the poor people, but also the prioritization of sustainability intervention by introducing and diffusing new technology in poor communities. Given the environmental problems that are associated with energy uses and economic growth in the poor areas, the priority is not only to secure people's energy needs, but also to improve the quality of energy services used by households and enterprises. For the two billion people who currently survive burning dung or wood for fuel, new renewable energy services[1] that are more sustainable are essential if the poor people are to escape from the economic and environmental traps of poverty (Mulugetta, Doig, Dunnett, Jackson, Khennas, & Rai, 2005). In fact, community development projects in most developing rural areas are implemented mainly through sustainable energy programs (Kadian & Kaushik, 2003). The challenge to decision makers is that much of what makes renewable energy technologies accepted and used by the poor people (Mulugetta et al., 2005).

Since the pioneering theoretical combination of psychoanalysis and behaviourism (Rutter, 1954), there has been a growing body of evidence and discussion linking social learning with improved mental and physical health, and other general well-being outcomes of community members (Berkes & Folke, 1998). Social learning depicts a process in which individuals learn from their community neighbours' previous experiences of adopting new technologies and the subsequent outcomes (Munshi, 2004). Focusing on the learning that occurs within a social context, social learning theory considers that people learn from one another, through such processes as observational learning, imitation, and modeling (Bandura, 1988). Webler, Kastenholz, and Renn (1995) address that social learning means more than merely individuals learning in a social situation. Rather, social learning envisions a community of people with diverse personal interests, but also common interests, who must come together to reach agreement on collective action to solve a mutual problem (Webler et al., 1995). Social learning theory suggests that results of people's behaviour have an impact on the motivation of their community members to engage in that behaviour (Rutter, 1954). Desiring positive results or effects, people will more likely to engage in a behaviour from which they expect a high probability of a positive outcome. Then, the behaviour is reinforced with positive outcomes and will lead other people to repeat the behaviour (Bandura, 1988). In

---

[1] Other than biomass in its traditional solid forms (wood and agricultural residues), renewable energy services that are more sustainable include biomass in non-traditional forms (converted into liquid and gaseous fuels, e.g. biogas), and solar, wind and hydro installations.

recent development practice, policy analysis, and management studies, scholars have proliferated the concept of social learning to comprise a collection of phenomena of learning by individuals through observation or interaction with their social context, learning by social aggregates, learning pertaining to social issues; and learning that results in recognizable social entities such as collective decision-making procedures, culture, etc (Maarleveld & Dangbegnon, 1999). Despite the growing recognition of advocating social learning in community development activities and policy agendas (Lee, 1993; Roling & Wagemakers, 1998; Ekins, 2004), however, there has been a paucity in literature so far addressing the social learning activities in community development projects, especially, in a context of a top-down planning system from the developing world.

This paper sets out some considerations from social learning theory that offers potentials of enhancing the diffusion of renewable energy technologies through active public participation in developing rural areas. The ideas are illustrated through a case study in a minority community that involves in the eco-village construction program in the middle mountainous area of southern China's Hainan province. Through the case study, I submit that farmers' cognition of future reinforcements in environmental soundness and economic savings can lead to active social learning and adoption of renewable energy technologies. I further conclude that it is practical to reinforce governmental top-down initiatives by encouraging social learning among stakeholders[2] of rural community development. Policy interventions for the purpose of sustainable community development should thus aim to enhance social learning and stakeholder collaboration.

THE CHINESE CONTEXT

Three sets of comments construct the context for the case study. The first concerns the top-down decision-making regime in China, the second concerns the energy problems in the developing rural areas of China, and the third is the *status quo* of its minority people that constitute the subject of the case study. Although some tentative efforts of decision-making have been made in terms of power decentralization, the political system of China is still very much top-down (Vernooy, Sun, & Xu, 2003). Policy decisions, and sometimes even more specific decisions, are made by the Communist Party Congress and are passed down to lower levels of governmental at provincial, city and local levels for specific implementation. In other words, critical decisions in China are made at the top of the political hierarchy and local jurisdictions are required to implement those decisions, although the directions of the final implementation may be subject to considerable local interpretation. Since the foundation of the People's Republic of China in 1949, the Chinese central government has realized the problems of rural energy security and has started to establish formal policies and plans to build legal capacity and frameworks (Figure 1), to encourage technical development and to support projects relating to the management of rural energy consumption (Yan, 1997; Zheng, Yang, & Shao, 2004). Following the direction of the plans developed by the central

[2] A stakeholder is defined here as "any group or individual who can affect or is affected by the achievement of an organization's purpose" (Freeman, 1984, p. 52). In community development projects, stakeholders include the private and public sectors that plan, manage, provide or deliver community development opportunities, such as government agencies, technical centres, and non-governmental organizations (NGOs), and those parties directly and indirectly affected by the projects, such as village cadres and residents.

government, government at the provincial and municipal levels have established its individual energy plans. The plans address the goal of improving rural energy system, but the plans are lack of appropriate implementation strategies and necessary communication among the local communities (Zhu, 2005). Additionally, energy plans at the local level are very scarce or vague in China.

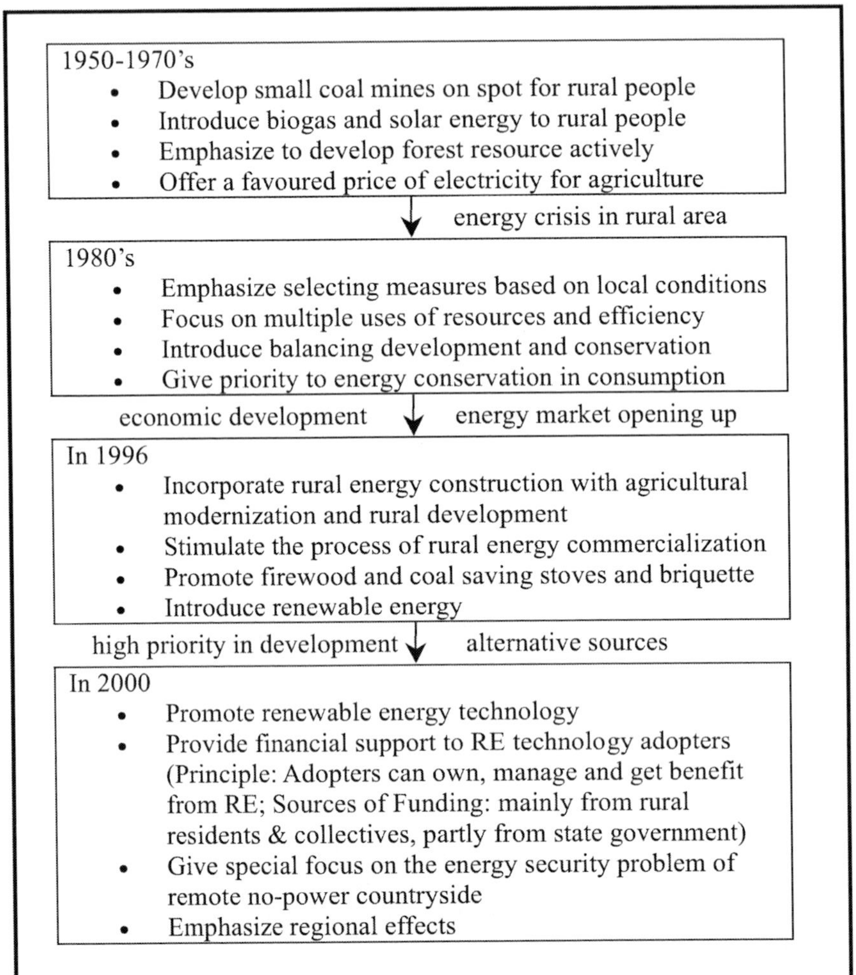

Figure 1. The Development of Rural Energy Policy in China
Adapted from: Zheng et al., 2004

China is a country with a large rural population[3] and significant power to produce agricultural products, which has grown rapidly with China's economic expansion in the 1980s (Zhu, 2005). Due to the generally undeveloped economic conditions along with China's large size and low density of railroads, highways, and power transmission lines, it is very difficult to ensure sufficient supplies of commercial energy (e.g. coal, oil, and electricity) to a large

[3] With rapid urbanization, the proportion of rural population in China reduced from about eighty percent in 1980's to 56 percent in 2006 and the total rural population of China in 2006 was 737 million (National Bureau of Statistics [NBS], 2006).

proportion of the rural areas (Wu, 1984). Hence, the Chinese rural people consume a considerable amount of energy, most of which is sourced from biomass energy substances, such as trees, firewood, and straw (Chen, 2003; Zhan & Yao, 2003). Owing to the lack of forest resources, peasants even have resorted to felling and digging up roots for burning in some extremely poor rural areas (Wu, 1984). In 2000, Chinese rural people consumed a total energy of 670 MTCE[4], almost equivalent to half of the total energy consumption in China (Zheng et al., 2004). As estimated, among the 1.5 billion people who rely on firewood to meet their cooking needs around the world, fifty percent of them live in Rural China (He, Jing, & Wang, 1992). The total biomass energy consumed in China each year accounts for approximately twenty percent of total annual global consumption, making China the largest consumer of biomass energy in the world (Zhu, 2005). The heavy dependence of Chinese rural people on biomass energy is influenced by levels of economic development, levels of technical knowledge, development and expertise, traditions, transportation infrastructure and the persistence of outdated approaches to energy consumption (Wu, 1984; Zheng et al., 2004; Zhu, 2005). The current major problems of energy utilization in Chinese rural areas include:

- Over-exploitation of forest resources and improper management of agricultural substances, which lead to resource waste and environmental pollution (Zhu, 2005);
- Local vegetation has been destroyed, resulting in severe water and soil losses (Wu, 1984);
- Low energy efficiency in rural families[5] (Wang, 2002; Wang, 2003; Zhu, 2005).
- Insufficient funds restrict technology introduction and diffusion in rural areas (Huang, 2002; Kang, 2002; Zhu, 2005).

The people in concern of the paper are from a Li community, one of the officially recognized ethnic minorities in Hainan province. The Li people in Hainan have a total population of more than 1 million, which occupies an important proportion of the approximate 8 million total population of the province. Most of the Li people live in the middle mountainous area of the province. Comparing to the Han majority people of the province, the Li people are much poorer on average and have less access to education in the remote areas where they live. The livelihoods of most Li people still depend on traditional subsistence agriculture in the mountainous areas and even in the periphery of the urban areas. Ever since the foundation of the People's Republic of China in 1949, people of ethnic minority groups have been regarded as backward, as they are incapable of managing their own affairs and very much in the position of needing to be improved and brought into the modernization process (Heberer, 1989; Gladney, 1994). Due to the lack of education, the Li people have been considered of being incapable of participating in decision-making and managing the local resources. Consequently, the only possible approach to their betterment has been considered to be through the top-down planning mechanism and the minority people

---

[4] MTCE stands for Mega Tonnes of Coal Equivalent. 1 kg Coal Equivalent = 7000 Kcal.

[5] If firewood is burned in a firewood-saving stove instead of the traditional open-fire stove, the thermal efficiency can be increased by twenty to thirty percent (Wang, 2003). If the agricultural substance and animal excreta can be utilized through anaerobic digestion process, the energy efficiency can be increased from twenty percent to more than sixty percent (Wang, 2002).

obey plans and policies constituted by the provincial government or the municipal regime which is established primarily by the majority Han people.

Objectives and Case Study Setting

This study aims to explore how social learning is incorporated in a governmental top-down initiative for promoting the adoption of renewable energy technologies or equipments in a developing minority community. By exploring who learns, what is learned, why it is learned, how it is learned and impacts of social learning occurring in the technology diffusion process of an eco-village in China's Hainan province (Figure 2), the study tries to shed some light on the potentials of social learning theory in stimulating renewable energy technology diffusion through active public participation in rural areas. Based on the study, this paper tries to provide a better understanding of social learning in community development projects for policy enhancement purpose. Although the author admits that a single case study may be insufficient to reveal the efficacy of social learning in promoting the adoption of renewable energy technologies or equipments, this approach is helpful for building on a rich set of empirical studies on rural community development in contemporary China.

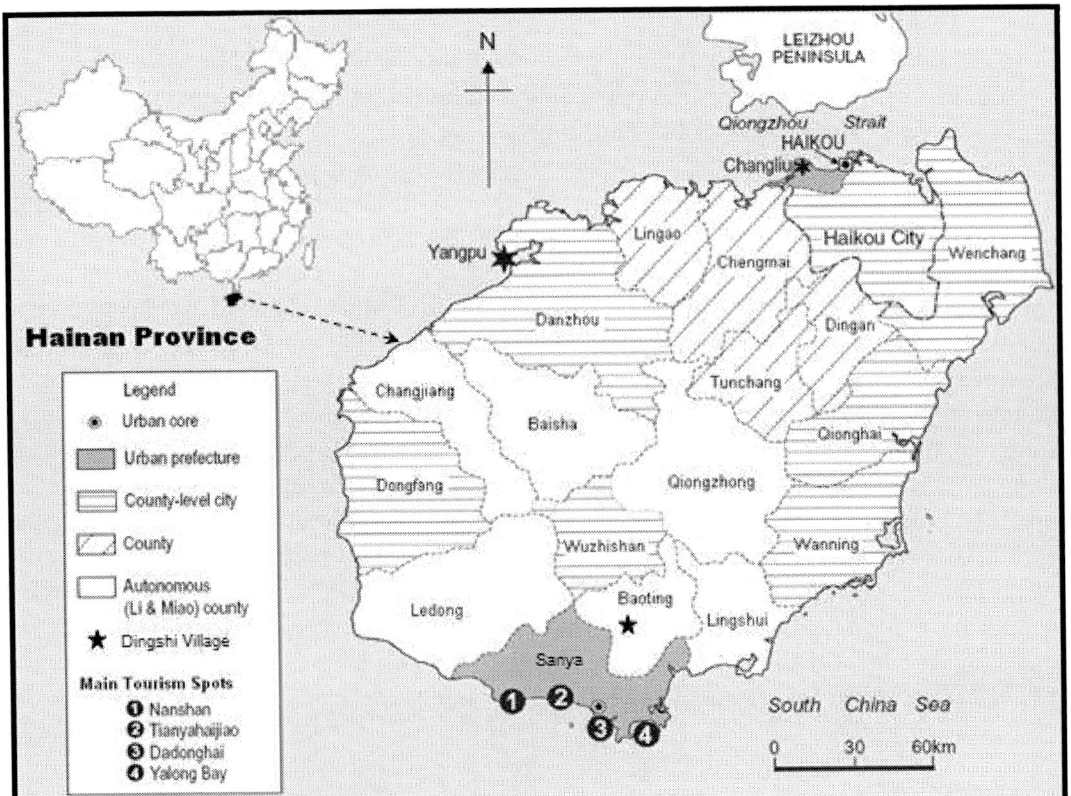

Figure 2. Location of Hainan Province
Source: Adapted based upon CAUPD 1988 and CAUPD 1999

This study focuses on an ongoing community development project in China's Hainan province, the *Civilized and Ecological Village* (CEV) program which started in 1999. Despite unprecedented fast rates of economic growth in rural China since the establishment of the country, perceptions of rural areas as poor, dirty and disordered exist among a majority of Chinese people (Shi, 1999; Wang, 2000). Paralleled to the pursuit of economic development, widespread environmental damage occurs in Hainan, although not as severe as that in mainland China, with the loss of mangrove areas and decreased forest coverage among the main environmental problems (Zhao, Lin, Liu, & Lin, 1999). In response to the global summon of Agenda 21 (United Nations Conference on Environment and Development [UNCED], 1992), the Hainan provincial government formulated an "*Eco-Province Strategy*" in 1999, aiming at building Hainan as the first "*Eco-Province*" in China (Liu, 2006). In 2001, the Hainan government formed the "*Civilized and Ecological Village*" Plan and encouraged rural people to build community capacity based on the existing rural settings as the *Local Agenda 21* (Liu, 2006). According to the goal of the provincial government, half of the natural villages in Hainan will have been turned into eco-villages[6] by the end of 2009 (Hainan Civilization-Ecology Office [HCEO], 2002). By the end of 2006, 6,523 villages (28.1 percent of the total natural villages in Hainan) had been renovated into eco-villages, which meets the expectation of the provincial government (Zhou, 2007).

The eco-village construction program in Hainan is illustrated through a case study in Dingshi village, which is a Li minority community located in the middle mountainous area of Baoting Li and Miao Autonomous County. In 2007, the village had 31 households and a total population of 151. The community economy is based on traditional agriculture and animal husbandry. Main products of the village include rubber, betel palm, coconut, lichee, bamboo, and cabbage palm. Before the eco-village program, energy source of the local residents in Dingshi, like the majority of the Chinese rural people, was firewood which was obtained mainly through deforestation activities. Coupled with the pressure from population growth, the objective of economic expansion in the village thrives for an ever-increasing need of energy sources which has generated intense pressure on the local resources. The village has been involved in the eco-village construction program since 2002 and major accomplishments have been achieved in terms of housing renovation, tapping water facility installation, power network upgrading, village road construction and adoption of renewable energy technologies and equipments, i.e. anaerobic digesters and solar water heaters. When renewable energy technologies are adopted by community residents gradually, the tension between economic growth and local resource protection is alleviated as a result of a more reasonable way of energy resource consumption and utilization. For example, as Bi and Haight (2007, p. 518) identify that,

> *"As anaerobic digester is incorporated into rural people's livelihood system, the technology can generate a number of benefits, including (1) offsetting energy needs; (2) reducing the reliance on chemical fertilizers while maintaining the nutrient cycle; (3) saving costs on commercial fuels and chemical fertilizers; (4) health condition improvements; (5) terminating deforestation activities and improving local*

[6] Takeuchi, Namiki, and Tanaka (1998, p.178) define an eco-village as "*a self-supporting area in which with the support of environmental conservation technologies, both a productive economy and the maintenance of semi-natural environmental systems can be realized.*"

environment; and (6) increasing the sustainability of local development more generally."

The author undertook the field research of the case study in fall 2007. First, demographic facts and information on Li culture were collected from all 31 households through interviews with the village director. Then, the author conducted an interview survey on residents' attitudes towards social learning in the eco-village program with an intended random sample of sixty adult residents. Following the pre-testing of the interview questionnaire, the author conducted face-to-face interviews with the residents in Dingshi village in Mandarin language, which can be fully understood and communicated in the community although it is not the primary language of the residents. The interview survey ended up with 42 usable interviews which represent seventy percent of the intended sample size. While doing interviews in the village, the author also conducted on-site observations of people's daily life in Dingshi, especially the operation of renewable energy technologies. In addition, eighteen key informant interviews were carried out with other three village leaders in the community as well as key governmental officials, technical experts, and Li scholars who were recommended and identified as stakeholders in rural community development by officers from Hainan Department of Lands, Environment and Resources [HLER]. All interviews were taped and transcribed with interviewees' permission.

Research Findings

Analysis of the discourses of local residents, village cadres, governmental officials, technical experts, and Li scholars aided the understanding of how social learning activities reinforced governmental initiatives of promoting renewable energy technologies in developing rural areas. The interviews revealed a complex range of economic and socio-political factors that had led to the success of adopting renewable energy technologies in Dingshi village. The case study also provides the practicability of encouraging social learning underlying the small-scale renewable energy technology diffusion campaign. In a context of developing rural areas where personnel and financial support are limited, implementation or enhancement of social learning may be able to generate more success towards community development projects. In the following sections, perspectives of different stakeholders in renewable energy technology diffusion in Dingshi village are presented separately.

Attitudes towards the Renewable Energy Program

Based on testimony of the farmers in Dingshi, it appears that both the cadres from the local collectives and officials from the Baoting municipal government have taken the leading role in organizing the eco-village construction program in the village since 2000, including the implementation of the renewable energy technology diffusion program. However, there were acknowledging narratives from the village leaders that it was the government of Zhaling Township, the governing regime of Dingshi, that initiated the campaign as a response to the "*conservation strategy for the middle mountainous areas of Hainan*" and the management of the village simply played an ancillary role in the community development projects. A provincial governmental official introduced that the provincial economic development strategy was designed based on the existing economic and environmental conditions of the whole province in late 1980s, when Hainan was officially designated as a province[7]. The justification of the "*conservation strategy*" for the middle area lies in its poor economic condition, low level of civilization, and high risk of environmental vulnerability. Specifically, the middle area has accumulated most of the tropical rainforest and seasonal forest of Hainan. Three major rivers of the province, including Nandu River, Changhua River and Wanquan River, originate from this area. As a result, the middle area is strategically important in preserving source water, conserving water and soil, purifying air, protecting biodiversity, ensuring ecological security, and maintaining ecological balance of the province. However, the natural forest in the middle of Hainan reduced by fourteen percent (about 83 thousand ha) from 1995 to 2005 due to agricultural expansion activities which originated from energy-related deforestation and poverty reduction struggles of the poor people. The quality of ecological functions of the middle area has thus reduced significantly. Given the importance of the ecological security of the middle area, the provincial government allege that promoting renewable energy technologies is a necessary content of the eco-village program, especially for villages located in the middle of Hainan.

The significance of rural environment and resources to economic development were commonly recognized by the village leaders of Dingshi. Although it is a top priority for the cadres to ensure economic growth in a community, the village leaders, as important nexus between government and farmers, understood and agreed the governmental policies of renewable energy program as an important content of eco-village construction. As they indicated:

> *"Without environmental resources, there is no agricultural development, not to mention economic growth";*
>
> *"Rural economic development should not be based on the costs of sacrificing rural environment and resources";*
>
> *"Since Dingshi involved in the eco-village construction program, the government of Zhaling Township has invited me to visit the Baoting Biogas Station as well as some model communities in the county. I was amazed by the convenience and efficiency of anaerobic digesters and solar water heaters. I agree and support the proposition of renewable energy program. I would like to be a pioneer of adopting renewable energy technologies in the village and I wish I could do my part to introduce the practice to the villagers."*

Despite the advocacy, both governmental officials and village cadres admitted the difficulty of sustainability intervention, especially for farmers whose economy and education were not ready for accepting new energy practice:

---

7 The northern and southern areas will be leading in economic expansion based on two main urban prefectures, Haikou and Sanya which have established comparably strong economic and industrial bases. The eastern and western parts of the province should be focusing on agricultural production, and industrial development will be less supported than the north and west. The provincial government has designated a "*conservation strategy*" for the middle mountainous areas, especially counties where minority people concentrate.

"With the help from Baoting rural technical centre, we tried to introduce anaerobic digesters and solar water heaters to our villagers. At the beginning, the new technologies seemed mysterious to the villagers and few of them cared about the new practices";

"At the beginning when we first introduced renewable energy technologies, villagers could not get why we triggered the intervention as they did not understand what was wrong with their traditional living styles which were passed on by their ancestors and had lasted for a very long time";

"The expenses to get necessary renewable energy equipment are definitely a considerable amount to the villagers. For example, when we first involved in the eco-village construction program, it took a household approximately 1,000 Yuan RMB to build an anaerobic digester. That money was a huge deal to most households, equivalent to half of the annual income of a household";

"Even there were a few villagers who showed interests when we first introduced the technologies, most of them shook head after knowing about the estimated expenses of construction".

These statements clearly show the difficulties of introducing new technologies to developing rural communities with farmers' unawareness of the new technologies as a major obstacle. Besides, poor economic condition has determined farmers' limited income to be spent on their livelihoods other than something they are not sure of. As farmers in Dingshi described their initial feelings about renewable energy technologies:

"When we started the eco-village construction program, we were aware that the village leaders were introducing some new energy technologies in the community on behalf of the government and the local rural technical centre. They presented that the technologies have promising benefits, but the practices are totally different from what we had been doing. As such, I did not want to be interrupted and I chose to keep my existing practice";

"At first, I was not interested in the renewable energy technologies which were introduced by the village leaders, as they showed up like some marketing persons trying to sell products for profit making purposes";

"Firewood, twigs and straws were the main sources of energy in my household. I had been using them for cooking since I could remember things. Before the renewable energy program, I felt nothing wrong with the traditional living style and so did other villagers, as all of them, even the village leaders themselves, had practiced in that way every day for a long time";

"They were introducing some sort of alternative sources of energy to us. At first, I felt that they were too expensive to afford. More importantly, we need to maintain the living of the entire family, buy chemical fertilizers for the plants, and finance children's tuition".

In spite of farmers' initial frustrating attitudes towards renewable energy technologies, the village leaders kept confidence and decided to be pioneers of adopting renewable energy technologies. A village cadre introduced that the anaerobic digesters and solar water heaters in the village leaders' households are among the earliest in the village. Meanwhile, farmers' initial attitudes towards the energy program had conveyed general information about difficulties and challenges in international development, which was captured by the Baoting municipal government. Accordingly, a financial incentive program was established by the government as a partial compensation for the costs of adopting renewable energy technologies in late 2001. Under the incentive, the villagers are encouraged to input initial labour and costs and adopt technologies first. After the practice being inspected by experts from technical centres, the users are eligible for governmental subsidy which usually covers one third of the total costs.

Coupled with the financial incentive program, the pioneer examples of anaerobic digesters and solar water heaters adopted by village cadres have attracted farmers' interests across Dingshi since 2002. As one village leader recalled, "*they became more proactive and showed more interest in the energy program than before, which was positive to the technology diffusion process*." The changing attitudes towards the energy program among the residents made it possible to diffuse the renewable energy technologies in the village.

Recognition of the Importance of Education and Training

Like most of Hainan's minority inhabitants, the Li people in Dingshi have a lowest average education level. A village cadre introduced that about forty percent of the villagers are illiterate and about 45 percent of the villagers have limited formal education equivalent to the level of primary school. Due to their isolation in mountainous areas, the Li people have rare access to quality education personnel from the Han majority, which has increased the difficulty of technology introduction activities in minority communities. Despite the constraints in rural education, officials from Baoting municipal government affirmed the importance of education and training in the rural area, especially where minority people concentrated. An official pointed out that, "*the current twin forces of nationality integration and technical change have increased the importance of education and training for minority people*". In accordance with the recognition, the government has tried to improve rural education in minority communities, as evidenced by the recent endeavours of establishing primary schools, regional technical support centres, and cultural exchange programs between Li and Han people. The efforts are improving the education of the Li people, but slowly. As manifested in Dingshi, the "*nine-year compulsory education*" program[8] still cannot be completely fulfilled by all children of appropriate ages. Parents in Dingshi have expressed a variety of reasons that lead to their caution and ambiguity on the "*nine year compulsory education*" program.

---

[8] In April 1986, the Chinese people's congress enacted "*The Compulsory Education Law of China*", which first legally enforced all children of appropriate ages in China to take six years of primary school and three years of secondary school education (In some places, the nine year compulsory education is carried out as a combination of five years' primary school and four years' secondary school education).

"Yes, I think children's education is important but it is too long for a kid to stay in school for nine years. Besides, we have too much agricultural work and we need more labour. I would prefer to have my children participate in the family agricultural work sooner";

"My children do not have very good academic records and I think the contents are too difficult for us minority people to understand. Moreover, the contents that children are learning in school are not as useful as we expect. At least, they are not useful in our agricultural work";

"Currently, many college and university students still have difficulty in getting a satisfied job after graduation, not to mention students with education lower than high school. The nine year compulsory education does not guarantee a job for the kids after graduation. It is better to for my children to work sooner than to waste time in school. To some extent, working is still a kind of learning, through which children can accumulate experiences for their future livelihoods";

"I appreciate the government's assistance in children's education in our village. However, the tuition for high school and above education is too high for us to afford".

Brainwashing through Technical Workshops

The technology diffusion program in Dingshi has stimulated villagers' passion of learning. A technical expert pointed out that, "*public education, in the technical context, is of prime importance in duplicating technical applications*". The expert explained that, owing to the rare opportunities of gaining access to the modern culture and science, the minority people have in depth faith of their traditional and existing living styles and it is challenging to alter their beliefs. Given the challenge, Baoting municipal government and Baoting Biogas Station have been organizing science and technology introduction workshops in minority communities. Themes of the workshop include renewable energy technologies as well as governmental environmental policies, case studies of achievements in model villages, and new agricultural practice. In Dingshi, villagers usually have opportunities to attend such sessions about four to six times a year. Being held during weekends in the cultural activity centre of the village, these workshops usually have good attendance. Since 2002, 26 workshops have been conducted in Dingshi. Villagers' awareness of education and training has been increasing with the ongoing workshops. Although ignorance of formal education still exists in a few households, the majority of villagers showed interests in the workshops.

Descriptive statistics of the responses to the question "*how many times have you participated in technical workshops*" shows that sixteen out of the 42 (38 percent) of the respondents have attended all 26 workshops, thirteen of them (31 percent) have attended most workshops, and 10 of them (24 percent) have attended about half of the workshops. Almost all the respondents expressed the willingness of participating in more workshops. A technical specialist who involved in the technical workshops in Dingshi told the researcher that, "*Many of the Li people in Dingshi learned at a slow speed in the workshops, but they still could be seen as trying hard and conscientiously. More importantly, many of them try to put what they learnt from the workshops into practice.*" To the majority of the villagers, the workshops exceed the meaning of traditional schools in that the workshops have provided communicating opportunities for farmers to enlarge vision and to share experiences of new applications:

> *"The workshops provide a window for us to know about how we can improve the quality of life with new energy technologies";*
>
> *"The workshops give us the vision of how an eco-village looks like. The technical specialists in the workshops provide guidance and specifics about how to construct our village into an eco-village as described by government officials. The technical workshops are very helpful";*
>
> *"In the workshops, experts not only confirm our accomplishments but also point out where we should improve. I think the cooperative work will lead to a better community";*
>
> *"My favourite section of the workshop is the question and answer session, during which I can get answers for most of my technical problems from the experts and my colleagues";*
>
> *"The attendees of the workshops in Dingshi are not just the local villagers. Also, I have seen people from other villages. It just shows that the workshops have significant impact on community development in the region".*

Villagers in Dingshi considered education and training in the energy program are necessary and important in eco-village construction and they expressed desirability of more technical-aid sessions in the village:

> *"The renewable energy technologies are completely new to us and I think we have much to catch up";*
>
> *"Interest in adopting the technologies does not work unless we have the know-how. Hence, I advocate the technical workshops and I would like to participate in the workshops as often as I can";*
>
> *"We wish we could have more opportunities to learn energy technologies and other applications suitable for rural residents";*
>
> *"The operation and maintenance of the practice also push us to learn continuously after the technologies were adopted".*

In Dingshi village, technical workshops not only delivered ideas from the government as how to construct a natural village into an eco-village, but also aroused residents' enthusiasm in technical training. However, it is incontestable that the workshops held in Dingshi were limited as only 26 workshops were conducted in seven years in the village and many villagers were not able to attend all the workshops. Under the circumstances, it is inevitable that villagers in Dingshi have to communicate with or learn from their neighbours about the new technologies. A governmental official introduced that, although the "*farmer-to-farmer extension*" policy is being promoted in rural Hainan as a strategy of promoting technology innovation and diffusion, the implementation of the policy is still at an early stage in terms of both time and scope, partly due to the shortage of technologists. Specifically, there are only three government-supported technical centres in Baoting Li and Miao Autonomous County. Although there are around 100 staff working in each centre, technicians specialize in different areas and those specialized in rural energy systems are very limited. Even in Baoting Biogas Station which focuses on rural energy systems, there are no more than twenty specialists who are capable of leading technical workshops in rural communities. Due to the lack of technical specialists, it is hard to conduct more technical workshops in rural Hainan.

LEARNING FROM NEIGHBOURS

As the "*farmer-to-farmer extension*" process occurred after several technical workshops in Dingshi, the renewable energy program in the village has flourished with increasing numbers of households who adopted renewable energy technologies. According to the statistics obtained from Baoting Biogas Station, all 31 households in Dingshi have adopted anaerobic digesters, and 20 households (65 percent) have installed solar water heaters by the summer of 2007 (Table 1). As is shown, the diffusion pace of anaerobic digestion has accelerated in Dingshi since 2003. Most anaerobic digesters (84 percent) were adopted between 2003 and 2006. Although eleven households involved in the survey have not adopted solar water heaters yet, many of them have expressed willingness of adoption in the near future. The municipal governmental officials have confirmed the effectiveness of technology diffusion in Dingshi. As one pointed out that:

> *"We were confident that the renewable energy program in Dingshi would have good result, but we did not expect that all the households had adopted anaerobic digesters by 2007. Dingshi has certainly established a solid step towards community sustainability, which has set up a good example for all communities involved in the renewable energy program. Clearly, the technical workshops were helpful for the residents. The cadres' pioneering examples were working well in demonstrating the benefits of the practice to the farmers. The subsidy we promised had stimulated the enthusiasm of the residents".*

Table 1. Renewable Energy Adoption Facts in Dingshi

	Households' Count & Percentage					
Year	2002	2003	2004	2005	2006	2007
Adopting Anaerobic Digester	3	4	7	9	6	2
Total adoption: 31	10%	13%	23%	29%	19%	6%
Adopting Solar Water Heater	1	2	6	4	3	4
Total adoption: 20	5%	10%	30%	20%	15%	20%

With regard to the energy program in Dingshi, the social learning process occurs as a combined studying experience from multiple sources, among which technical workshops and the *"farmer-to-farmer extension"* process are the major sources of learning (Table 2). Since the learning experience in the village occurs continuously, it was difficult for the villagers to articulate how or where they learned a specific knowledge. Data analysis results of the questionnaire survey reveal that a lot of villagers in Dingshi actually learned a lot of knowledge from both technical workshops and their neighbours, e.g. general governmental policies/plans, general scientific/technical knowledge, benefits and rationale of anaerobic digestion. Some villagers also responded other sources of learning, including TV programs, radio broadcasting, VCD/DVD, and their out-of-village relatives.

Table 2. Contents of Social Learning in Dingshi

	Community Residents (N = 42)			
	Technical Workshops		Farmer-to-farmer	
Contents	# of responses	% of responses	# of responses	% of responses
Overall policies/plans	39	93	5	12
Case studies of model villages	32	76	2	5
Potential areas of improvement	33	79	7	17
Resource/environmental protection	24	57	18	43
General scientific/technical knowledge	28	67	23	55
Source of funding	33	79	7	17
Agricultural practice	25	60	15	36
Anaerobic digester				
Benefits	33	79	12	29
Rationale	27	64	25	60
Construction	11	26	31	74
sources of raw materials	9	21	33	79
operation/maintenance	7	17	35	83
security issues	23	55	19	45
uses of digester sludge	5	12	37	88
Solar water heater				
Benefits	16	38	23	55
market information	5	12	19	45
Installation	2	5	26	62
operation/maintenance	2	5	25	60
security issues	13	31	14	33
Other rural energy technologies	15	36	2	5

Survey results indicate that both the technical workshops and the *"farmer-to-farmer extension"* process are important components of the social learning process in Dingshi village. Most of the time, the two processes complement each other. Workshops are major tunnels of delivering general governmental policies/plans, model villages, information about areas to be improved, and benefits/rationale of anaerobic digesters. A governmental official told the researcher that holding technical workshops in more villages is one of the most important tasks in the eco-village program of Hainan because workshops enable the township government to prioritize local collective consciousness by directly communicating with the villagers and ideas between both sides can be exchanged effectively and efficiently in discussions. While workshops facilitate the communication between government and villagers, the "*farmer-to-farmer extension*" process is important in turning ideas into practice. As shown in Table 2, many villagers in Dingshi learned how to construct, operate, and maintain their anaerobic digesters and solar water heaters more from their neighbours than from the workshops. Learning activities in the "*farmer-to-farmer extension*" process involve continuous observing, taking action, monitoring usage and adjusting practice. Learning from neighbours assures that villagers can get timely responses and answers of technical questions when they need them, which to some extent reduces the risks of abandoning devices out of order. According to the interviews with government at different levels, some experimental projects of renewable energy were conducted in rural Hainan as early as early 1980s, many projects failed of continuous uses soon after they were established because many users could not get timely responses to solve the technical problems of the devices. With the benefits of continuously using the renewable energy technologies, villagers in Dingshi showed strong willingness to share their experiences and knowledge with their neighbours and would like to provide their colleagues with assistance on the application of renewable energy technologies.

Attitudes towards Stakeholder Collaboration

The renewable energy program in Dingshi has involved collective action between local authority and the public. Villagers, village cadres, technical experts, and government had collectively participated in the program, demonstrating that community development projects are collaborative works with different stakeholders' participation as the key. Before the renewable energy program, villagers in Dingshi, like most Li minority people, were lack of a tradition of mutual-aid cooperation with local authorities and technical centres, mainly because their subsistence mode of agricultural production did not require a network to share collective intelligence on community development. As a much isolated minority group, Li people are generally not interested in politics and they are afraid of contacting with government officials. Most of the time, minority people only adhere to government decisions, not to mention participation in decision-making activities. The social learning on renewable energy technologies in Dingshi requires and stimulates the minority residents to participate, which has grounded the basis for stakeholder collaboration in the village.

Effective participation requires an environment of trust, transparency, respect, and openness, which take time to develop, especially in minority communities where people are not used to working together. Stakeholders of the renewable energy program in Dingshi generally acknowledged that participants' interaction and cooperation which occur in the social learning process are vital to the implementation of the renewable energy program. Despite the different roles of stakeholders in the program, participants share a common objective—sustainable community development in Dingshi. Various interviewees of the study showed strong support of stakeholder collaboration in community development projects for active collaboration has led to significant changes in Dingshi in terms of energy consumption, living conditions, public environmental awareness, and overall appearance of the village. Government interviewees were satisfied with the changes and affirmed the strategy of strengthening stakeholder collaboration in rural community development projects, because collaboration in minority areas will not only strengthen the communication between decision makers and the public but also reconcile the ideals of the Chinese nationalism with ethnic diversity. Most importantly, the collaboration will motivate rural people positively towards economic development and environmental soundness. Technical experts affirmed the effectiveness of social learning in technology diffusion in developing communities. They also recognized the social learning process as a good opportunity to learn traditional knowledge of local residents, which in turn contributes to their work. As an official of Baoting Biogas Station noted:

> "*Due to the fear of the majority Han Chinese, Li people have a long tradition of avoid communicating with the Han people. It is very difficult to understand the Li culture, custom and traditions which may challenge the implementation of governmental policies. Being technologists, we support and introduce knowledge of the dominant society to less developed communities as a mission. Most importantly, we must acknowledge aboriginal traditions. As we all have brainstorming in the technical workshops, we also have chances to understand the Li people so that we can integrate local knowledge with science and technology.*"

Village cadres of Dingshi are proud of the achievements of the village since the turn of the century and they emphasized that it is important to foster interaction of stakeholders, manage conflicts, and encourage developing mutually beneficial solutions among stakeholders. They also noted the managerial function of stakeholder collaboration in terms of facilitating conflict resolution among villagers. With the participation in the community development project, villagers in Dingshi have more opportunities to communicate with the authorities as well as their colleagues. Unmistakably, they can seek discussion opportunities to resolve conflicts that they may have. As expected, villager interviewees concerned stakeholder cooperation very important in community development projects and would continue participating in programs alike. Given such perspectives, it is apparent that the minority people in Dingshi recognize the positive changes of the renewable energy program in the village and thus maintain interests and trust in the community development projects. They are becoming more open and social in the projects notwithstanding their traditional tendency of being uncommunicative. This explains, in part, the facts of the fast technology diffusion rates in Dingshi, and leads to another area of concern—the potential of strengthening the strategy of encouraging social learning and stakeholder collaboration across rural Hainan.

DISCUSSION

Given China's one-party political system, decisions of the Chinese central government are implemented as obeisance of the directives at lower levels. Likewise, the procedures of promoting environmental policies and implementing environmental intervention are highly centralized and depend overwhelmingly on the execution of government at lower levels as well as the participation of local people. Numerous scholars have criticized the top-down planning system and suggest that decentralized governments are not only more efficient and less corrupt but also can promote economic development and democracy better than centralized ones (Wunsch, 1999; Oyugi, 2000, Wunsch, 2001; Kohl, 2003, etc.). However, the collective responses of farmers in Dingshi to the provincial environmental policies may auspice that an adaptive approach in governance is necessary for communities at different levels of development. In the majority of minority communities in China, like Dingshi, where people are isolated and generally lack of education, a top-down planning system is more appropriate than a bottom-up one. A top-down planning system does not mean coercion. Good governance, which is central to development practice, is not only about making policies but is also about the creation of creative partnerships and communication between the state and other actors. This occurred in the Dingshi case to a certain degree in that governmental environmental policies were implemented and enhanced through social learning process with the participation of multiple stakeholders. It implies that environmental intervention consisting only of strict plans is insufficient to be fully ensured in less developed communities. Rather, implementation is a key concern of the environmental policies. It is impractical to expect community residents to respond to the plans without proper instruction or guidance. With an attempt to find different ways to introduce technical information, communication tools such as workshops, community meetings and discussion forums will make the policy implementation process more accessible to a wide range of stakeholders, especially the grassroots.

It becomes apparent in the case of Dingshi that innovative leaders are important and useful to introducing and diffusing small rural technologies in less developed communities. Community leadership is different from governmental governance—community leaders cannot get their jobs done by only relying on power and formal authority. As identified by Pigg (1999), community leaders must rely on networks and influence, and specifically the relationships developed through extensive interaction within the community. Effective community leaders assist in developing important relationships, establishing communication and imparting community direction, therefore providing important link between the authority and the grassroots (Ricketts & Ladewig, 2008). This requires a community leader to have not only strong management skills but also communicative abilities. Often, community leadership works towards the common good of the community, which utilizes community-wide action for addressing mutual problems (Ricketts & Ladewig, 2008). In the context of technology diffusion, village cadres can be viewed as not only pioneers of adopting new technologies but also possible extensionists of the technologies, as community leadership aims at inspiring and mobilizing others to undertake collective action in pursuit of the common good (Greenleaf, 1996). Since village cadres have the potential to trigger the "*farmer-to-farmer extension*" process in their communities, it is promising to give priority in training village leaders as extensionists of new technologies and practices. Model projects accomplished by the village cadres can demonstrate possible benefits of the new technology or practice to community members, which may be able to nurture a community-based multiplier effect of improving rural practices.

Stakeholders' participation is critical in the social learning process. If the potential of social learning is to be maximized, skilled planning and facilitation for participation are needed. Rather than making sermon in communities, decision makers should be more active in stimulate the interaction of stakeholders, resolve possible conflicts, and encourage the development of mutually beneficial projects. As in Dingshi, what really motivated local farmers to participate in the community development projects was the recognition of the potential benefits they would obtain after adopting renewable energy technologies. Residents' energy consciousness, including willingness to adopt new energy devices, was built upon a cumulative learning process with exposure to different sources of information and feedback. Participation and social learning in community development projects involve continuous activities, including observing, taking action, monitoring usage and adjusting practice in their environment. Although the findings reveal improved skills and awareness among the residents and some positive changes of environmental and resource practices in Dingshi, much more efforts are needed to stimulate the interaction among stakeholders and facilitate participation in community development projects. This indication may be attributed to the less developed nature of the community as well as the conservative characteristics of the rural people, which are not exclusive in the Dingshi case but part of the general developing world. As evident as the residents in Dingshi, rural people can be aware of environmental soundness and economic savings of energy consumption with proper training. Given the common challenges of limited funding and personnel in developing world, it is woolgathering to send technologists to communities where there is a need. Alternatively, it is feasible to recruit volunteers in colleges and universities and have them introduce governmental decisions and policies to rural people. To follow up, continuous interaction between the authority and the grassroots are certainly needed.

Community development projects in developing rural areas often struggle with the challenge of involving the public, as the poor often presume that they are intervened by the authority or at least persuaded to cooperate on a conservation project. A compelling issue arising in the village of Dingshi is that the township government pre-selected the village as a target of introducing renewable energy technologies. As a pre-condition, the availability of technical specialists allowed the villagers to have the preliminary perception of the development or aid program. However, it is hard to ensure such technical workshops in most natural villages yet to be selected and built as an eco-village. Although some residents of Dingshi did respond other sources of learning, including TV programs, radio broadcasting, VCD/DVD, and their out-of-village relatives, it is unrealistic to expect all rural poor people to learn by themselves spontaneously or to participate in community development projects without face-to-face communications. Therefore, it is critical to provide more technical workshops in rural communities. As the rural people have more access to information and sources of technologies, they will have more chances to learn unaided, evaluate information, invent solutions to problems and share knowledge with others. Rural Hainan, especially minority people concentrated area, is in need of a supportive strategy to deliver education and training programs in a more effective manner to the poor people. Such programs should be provided on a regular basis, not limited to renewable energy technologies but broad community development initiatives, such as community-based industries, advanced agricultural practice, and plantation of economic crops for after all, both the authority and the public are strongly oriented towards economic development.

CONCLUSIONS

This article focuses on a rural community which involves in local community development projects initiated by the government of China's Hainan province, exemplifying the process and roles of various stakeholders involved in active social learning activities which have spurred the diffusion of renewable energy technologies in the community. Based on the definitions of social learning in literature, the ongoing learning activities in the community development projects of Dingshi village are positive—various stakeholders involve in continuous learning actively, a top-down governmental initiative is implemented effectively, and actors share the benefits of their accomplishments in adopting renewable energy technologies. As supported by the government of Hainan, technical workshops have grounded the foundation for public participation in community social learning and have enhanced the communication between authority and community residents. Although the technical workshops in Dingshi are limited in terms of number and frequency, they are effective in illustrating the big policy picture and rationale of renewable energy technologies to the public. With participation in the technical workshops, community residents are able to identify the potential benefits of renewable energy technologies. This motivates villagers to participate in community development projects and to learn from all sources available, especially their neighbours. Given the context of developing communities where economic and education levels are low, it is challenging for the residents to collectively respond to environmental policies without proper instruction and guidance. Consequently, I submit that technical workshops are important and necessary social learning activities of development projects in the developing world, no matter what the objective of the community in question is—a development or aid intervention, or is to be persuaded to cooperate on a conservation project. With proper knowledge of the intervention program, community residents can learn from their neighbours and possibly duplicate the new practice, which assures the implementation of development initiatives. Should the community development projects be of financial input, subsidies or other financial incentives for the participants might be able to stimulate the progress of the projects.

The results of the study confirm Bandura's (1988) contention that social learning behaviours can be reinforced with preliminary positive outcomes and will lead people to repeat the behaviours; in this case, active social learning can lead to positive community development outcomes. Numerous quotations in the study indicate that poor people in less developed communities are willing to engage in learning and adopting new practices with perceived benefits of improving their livelihoods, even though they remain equivocal about the necessity of formal education. However, the major conclusion varies from the prevailing belief of active local participation in decision-making as a precondition for benefits to reach communities (Renn, Webler, & Wiedemann, 1995; Plein, Green, & Williams, 1998; Smith & McDonough, 2001; Tabbush, 2004; etc.). Based on the study, I argue that in a top-down planning system, participatory development in developing communities is achievable through governmental intervention and active social learning, without which neither the authority nor the community residents would be able to receive benefits from community development projects. Depending on the level of civilization, empowering the grassroots in decision-making is neither the only way to ensure local community to receive the benefits of participation nor an ultimate goal in itself, especially in developing rural areas. Here is why I further submit that community development programs for sustainability intervention are necessary for developing rural areas with identified problems, yet the implementation of the programs should be flexible and adaptive with encouraging active social learning as the executing basis.

REFERENCES

Bandura, A. (1988). Social Foundations of Thought and Action: *A Social Cognitive Theory.* Englewood Cliffs: Prentice Hall.

Berkes, F. & Folke, C. (1998). *Linking Social and Ecological Systems: Management Practices and Social Mechanisms for Building Resilience.* Cambridge: Cambridge University Press.

Bi, L. & Haight, M. (2007). Anaerobic digestion and community development: a case study from Hainan Province, China. *Environment, Development and Sustainability, 9(4),* 501-521.

CAUPD. (1988). *The Comprehensive Land-use Plan for Haikou.* Beijing: China Academy of Urban Planning and Design.

CAUPD. (1999). *The Comprehensive Land-use Plan for Sanya.* Beijing: China Academy of Urban Planning and Design.

Chen, J. B. (2003). Research on foreground for renewable energy as an industry in China. *Renewable Energy, 3,* 10-12.

Dincer, I. (2000). Renewable energy and sustainable development: a crucial review. *Renewable & sustainable energy reviews, 4(2),* 157-175.

Ekins, P. (2004). Step changes for decarbonising the energy system: research needs for renewables, energy efficiency and nuclear power. *Energy Policy, 32(17),* 1891-1904.

Freeman, R. E. (1984). *Strategic Management: A Stakeholder Approach.* Boston: Pitman.

Gladney, D. C. (1994). Representing nationality in China: refiguring majority/minority identities. *The Journal of Asian Studies, 53,* 92-123.

Goldemberg, J. (1995). Energy needs in developing countries and sustainability. *Science, 269(5227),* 1058-1059.

Greenleaf, R. K. (1996). *On Becoming a Servant Leader.* San Francisco: Jossey-Bass.

HCEO [Hainan Civilization-Ecology Office]. (2002). *The Construction Plan of Civilized and Ecological Villages in Hainan.* Haikou: The Chinese Communist Party Committee of Hainan Province.

He, Q., Jing, W. Y. & Wang, Y. T. (1992). *Introduction of Environmental Studies.* Beijing: Tsinghua University Pres.

Heberer, T. (1989). *China and Its National Minorities: Autonomy or Assimilation?* Armonk: ME Sharpe.

Huang, Y. (2002, October 22). To change straws to manure in form of animal excrement: an effective prescription to exterminate the phenomenon of burning. *China Environment News.*

Kadian, P. & Kaushik, S. (2003). *Rural Energy for Sustainable Development: Participatory Assessment of Energy Resources.* New Delhi: Deep & Deep Publications Ltd.

Kang, R. (2002, September 24). Prohibition mechanism of straw burning should be put in order. *China Environment News.*

Kohl, B. (2003). Democratizing decentralization in Bolivia: the law of popular participation. *Journal of Planning Education and Research, 23(2),* 153-164.

Lee, K. (1993). *Compass and Gyroscope: Integrating Science and Politics for the Environment.* Washington D.C.: Island Press.

Liu, S. Z. (2006). *The Tutorial of Eco-village Construction.* Haikou: Hainan Press.

Maarleveld, M. & Dangbegnon, C. (1999). Managing natural resources: a social learning perspective. *Agriculture and Human Values, 16(3),* 267-280.

Mink, S. D. (1993). *Poverty, Population, and the Environment.* Washington D.C.: The World Bank.

Mulugetta, Y., Doig, A., Dunnett, S., Jackson, T., Khennas, S. & Rai, K. (2005). *Energy for Rural Livelihoods: A Framework for Sustainable Decision Making.* Warwickshire: ITDG Publishing.

Munshi, K. (2004). Social learning in a heterogeneous population: technology diffusion in the Indian Green Revolution. *Journal of Development Economics, 73(1),* 185-213.

NBS. (2006). *China Statistics Abstract 2006.* Beijing: China Statistics Press.

Oyugi, W. O. (2000). Decentralization for good governance and development: the unending debate. *Regional Development Dialogue, 21(1)*, 3-22.

Pearce, D. W. & Warford, J. J. (1993). *World without End: Economics, Environment, and Sustainable Development.* New York: Oxford University Press.

Pigg, K. E. (1999). Community leadership and community theory: a practical synthesis. *Journal of the Community Development Society, 30(2),* 196-212.

Plein, L. C., Green, K. E. & Williams, D. G. (1998). Organic planning: a new approach to public participation in local governance. *The Social Science Journal, 35(4)*, 509-523.

Posa, M. R. C., Diesmos, A. C., Sodhi, N. S. & Brooks, T. M. (2008). Hope for threatened tropical biodiversity: lessons from the Philippines. *BioScience, 58(3),* 231-240.

Renn, O., Webler, T. & Wiedemann, P. (1995). *Fairness and Competence in Citizen Participation. Evaluating Models from Environmental Discourse*. Dordrecht: Kluwer Academic Publications.

Ricketts, K. G. & Ladewig, H. (2008). A path analysis of community leadership within viable rural communities in Florida. *Leadership, 4(2),* 137-157.

Roling, N. & Wagemakers, A. (1998). *Facilitating Sustainable Agriculture: Participatory Learning and Adaptive Management in Times of Environmental Uncertainty.* Cambridge: Cambridge University Press.

Rutter, J. B. (1954). *Social Learning and Clinical Psychology.* Englewood Cliffs: Prentice Hall.

Shi, T. (1999). Economic development and village elections in rural China. *Journal of Contemporary China, 8(22),* 425-442.

Smith, P. D. & McDonough, M. H. (2001). Beyond public participation: fairness in natural resource decision-making. *Society and Natural Resources, 14(3),* 239-249.

Tabbush, P. (2004). Public money for public good? Public participation in forest planning. *Forestry, 77(2),* 145-156.

UNCED [United Nations Conference on Environment and Development]. (1992). *Agenda 21, the United Nations Programme of Action from Rio.* London: UN Department of Public Information.

Vernooy, R., Sun, Q. & Xu, J. C. (2003). *Voices for Change: Participatory Monitoring and Evaluation in China.* Kunming: Science & Technology Press.

Wang, C. C. (2003). Classification of new energy and renewable energy. *Solar Energy, 1,* 14-15.

Wang, Q. Y. (2002). Current status of China's renewable energy and existing obstacles and countermeasures. *Energy of China, 7,* 39.

Wang, X. H. & Feng, Z. (2002). Sustainable development of rural energy and its appraising system in China. *Renewable & sustainable energy reviews, 6(4),* 395-404.

Wang, Z. (2000). Poverty eradication: the role of the China society for promoting the Guangcai Programme. *Regional Development Dialogue, 21(2),* 158-162.

Webler, T., Kastenholz, H. & Renn, O. (1995). Public participation in impact assessment: a social learning perspective. *Environmental Impact Assessment Review, 15(5),* 443-463.

Wu, K. (1984). Energy sources in China - present and future. In V. Baum (Ed.), *Energy Planning in Developing Countries.* New York: Oxford University Press.

Wunsch, J. S. (1999). Institutional analysis and decentralization: developing an analytical framework for effective Third World administrative reform. *Public Administration, 11(5),* 431-451.

Wunsch, J. S. (2001). Decentralization, local governance and "recentralization" in Africa. *Public Administration and Development, 21(2),* 277-288.

Yan, C. L. (1997). *Report on China Energy Development.* Beijing: Economic Management Press.

Zhan, H. & Yao, S. H. (2003). Some speculation about general status of energy and development in China. *Energy Engineering, 3,* 1-4.

Zhao, L., Lin, Y., Liu, S. Z. & Lin, H. F. (1999). *The Research Report on the Agricultural Resources Use and Sustainable Development in Hainan.* Haikou: The Chinese Communist Party School of Hainan Provincial Committee.

Zheng, Y. S., Yang, M. Y. & Shao, Z. (2004). *Rural Energy Policy in China.* Beijing: Research Group of Energy Policy, Institute of Quantitative and Technical Economics, Chinese Academy of Social Sciences.

Zhou, W. Z. (2007). Furthering the progress of Eco-village construction in Hainan. *The Spiritual Civilization Construction of Hainan, 68,* 8-23.

Zhu, J. G. (2005). Agricultural renewable energy and its management in China. In R. Bradbrook, R. L. Lyster, R. L. Ottinger & W. Xi (Eds.), *The Law of Energy for Sustainable Development.* New York: Cambridge University Press.

In: Energy Costs, International Developments... ISBN: 978-1-60741-700-2
Ed: Leszek Kowalczyk and Jakub Piotrowski

Chapter 4

ENERGY AND SUSTAINABLE DEVELOPMENT IN SUDAN

Abdeen Mustafa Omer
17 Juniper Court, Forest Road West, Nottingham NG7 4EU, UK

ABSTRACT

Growing concerns about social and environmental sustainability have led to increased interest in planning for the energy utility sector because of its large resource requirements and production of emissions. A number of conflicting trends combine to make the energy sector a major concern, even though a clear definition of how to measure progress toward sustainability is lacking. These trends include imminent competition in the electricity industry, global climate changes, expected long-term growth in population and pressure to balance living standards (including per capital energy consumption). Designing and implementing a sustainable energy sector will be a key element of defining and creating a sustainable society. In the electricity industry, the question of strategic planning for sustainability seems to conflict with the shorter time horizons associated with market forces as deregulation replaces vertical integration. Sudan is an agricultural country with fertile land, plenty of water resources, livestock, forestry resources, and agricultural residues. Energy is one of the key factors for the development of national economies in Sudan. An overview of the energy situation in Sudan is introduced with reference to the end uses and regional distribution. Energy sources are divided into two main types; conventional energy (woody biomass, petroleum products, and electricity); and non-conventional energy (solar, wind, hydro, etc.). Sudan enjoys a relatively high abundance of sunshine, solar radiation, moderate wind speeds, hydro, and biomass energy resources. Application of new and renewable sources of energy available in Sudan is now a major issue in the future energy strategic planning for the alternative to the fossil conventional energy to provide part of the local energy demand. Sudan is an important case study in the context of renewable energy. It has a long history of meeting its energy needs through renewables. Sudan's renewables portfolio is broad and diverse, due in part to the country's wide range of climates and landscapes. Like many of the African leaders in renewable energy utilisation, Sudan has a well-defined commitment to continue research, development, and implementation of new technologies. Sustainable low-carbon energy scenarios for the new century emphasise the untapped potential of renewable energy resources are required. Rural areas of Sudan can benefit from this transition. The increased availability of reliable and efficient energy services stimulates new development alternatives. It is concluded that renewable environmentally friendly energy must be encouraged, promoted, implemented, and demonstrated by full-scale plan especially for use in remote rural areas.

Keywords: Sudan, energy, impacts on environment, sustainable development, mitigations.

1. Introduction

Climate change is a hot issue in world politics. The use of fossil fuel is seen as a cause of critical global warming. Long-term energy options currently considered include the petroleum, electricity, and biomass. Sustainability is increasingly becoming an element of world politics, although there is not yet agreement on a clear definition and indicators are still not yet fully agreed upon that would effectively enable the establishment of the sustainable development, which is so eagerly sought. However, the snowball is rolling: the process of designing sustainable development has started, and it is only a matter of political will, negotiations and time for it to accelerate, hopefully in the right direction. The job is tough, and the variety of stakeholders involved in today's globalisation process makes the whole story even more interesting and challenging. And yet, globalisation and sustainable development are bound to become tautological. The goal of sustainability has different meanings and measures specific to various regions of the planet, each with their own economics, histories, and cultures. A pathway to sustainable development that is reasonable, achievable and hence natural in one country.

In developed countries, most investments in electricity generation have paid back their initial capital costs. Research and development of new electricity generation technologies are well under way, and these technologies hold good promise of achieving commercial feasibility. New investments in electricity generation have not been aggressively pursued in recent years. Rather, policies for existing energy infrastructures have included improved options for more sustainable electricity generation, e.g., less pollution, higher fuel efficiency and life extension. However, are, at best, only interim solutions. Stakeholders interaction is essential to create a culture of sustainability, with educational, regulatory, economic, environmental and ethical dimensions that can have an impact on society overall. Such a culture of sustainability must be designed, managed and measured in ways compatible with societal attitudes towards risk, including changes in perspective over medium and long-term time horizons. Risk-related concepts play a large part in what technologies the public views as sustainable, whether it concerns potentially large accidents from electric generation technologies, greenhouse gas emissions (GHGs), vulnerability to natural disasters, or decommissioning and site reclamation problems. Research and development costs and expected lead times related to electricity generation, also play a role in stakeholder positions and public attitudes [1].

There are increasing concerns but the sustainability of the energy sector, ranging from impacts of current operation to the choice of future options for system development. These concerns include such issues as health and safety, environmental emissions, use of energy and materials resources, regulated versus competitive markets, vulnerability of electrical energy networks, cost and equity issues among users of diverse size, and appropriate technology for the development and commercialisation of improved supply and end-use equipment. Such concerns span the geographic spectrum from local issues such as cost and siting, to regional issues like acid rain [2], state and national issues such as deregulation and social acceptance of competing technologies, and global issues such as climate change [3]. Decision-makers and planners in the energy sector must address these issues, and fit them into a broader framework of national and world energy and develop policies. The methodological framework needed to assist decision-makers can be generalised in part, because most of these problems share common elements, and can be characterised by their combination of: (1) complexity (2) dispersed solutions (3) finite resources, and (4) societal impacts.

The move towards a de-carbonised world, driven partly by climate science and partly by the business opportunities it offers, will need the promotion of environmentally friendly alternatives, if an acceptable stabilisation level of atmospheric carbon dioxide is to be achieved. This requires the harnessing and use of natural resources that produce no air pollution or greenhouse gases and provides comfortable coexistence of human, livestock, and plants. This study reviews the energy-using technologies based on natural resources, which are available to and applicable in the farming industry. Among these are greenhouses, which are necessary for the growth of some plants (i.e., vegetables, flowers, etc.) in severe climates. Globally, buildings are responsible for approximately 40% of the total world annual energy consumption. Most of this energy is for the provision of lighting, heating, cooling, and air conditioning. Increasing awareness of the environmental impact of CO_2, NO_x and CFCs emissions triggered a renewed interest in environmentally friendly cooling, and heating technologies. Under the 1997 Montreal Protocol, governments agreed to phase out chemicals used as refrigerants that have the potential to destroy stratospheric ozone. It was therefore considered desirable to reduce energy consumption and decrease the rate of depletion of world energy reserves and pollution of the environment.

In the areas of power systems planning, one of the most pressing problems that exemplifiers this combination of characteristics is the intersection between:

a) The rapid liberalisation of energy industries driving short-term actions to maximise stakeholder values, and
b) The possible restrictions on greenhouse gas emissions proposed to meet the problem of global climate change.

The National Electricity Corporation (NEC) has the goal of improving electricity system network in order to achieve sustainability. Within this broad aim, this search has specific goals, including:

- To involve a wide-range of electricity sector participants, including utilities, regulators, environmentalists, and customers.
- To identify and implement a wider range of measures that will be direct indicators of sustainability than the current response impacts and pollutant outputs.
- To create improved electricity sector modelling tools that simulate electricity system operation under competition, and include transmission and distribution effects more directly into strategic planning.
- To consider technology options beyond the current generation.

- To increase the comprehensiveness of current life-cycle assessment, by including broader range of technologies, inputs and outputs are sensitive to technology choice versus the state of underlying production and transport infrastructures.
- To improve decision analysis tools to assist stakeholders in reaching consensus on complex alternatives, option portfolios, and flexible contingencies.
- To make the data and models developed for understanding the relationship between power systems and sustainability, available and usable through internet access.

NEC handled wide variety of electricity generation technologies (Table 1), including:

- Thermal (fossil, combined cycle, and combustion turbines).
- Storage (pumped hydro, batteries, and compressed air).
- Non-dispatchable technologies (solar, wind, cogeneration, and load management).

The data required to perform the trade-off analysis simulation can be classified according to the divisions given in Table 2: the overall system or individual plants, and the existing situation or future development.

2. Environmental Aspects

Environmental pollution is a major problem facing all nations of the world. People have caused air pollution since they learned to use fire, but man-made air pollution (anthropogenic air pollution) has rapidly increased since industrialisation began. Many volatile organic compounds and trace metals are emitted into the atmosphere by human activities. The pollutants emitted into the atmosphere do not remain confined to the area near the source of emission or to the local environment, and can be transported over long distances, and create regional and global environmental problems.

A great challenge facing the global community today is to make the industrial economy more like the biosphere, that is, to make it a more closed system. This would save energy, reduce waste and pollution, and reduce costs. In short, it would enhance sustainability. Often, it is technically feasible to recycle waste in one of several different ways. For some wastes there are powerful arguments for incineration with energy recovery, rather than material recycling. Cleaner production approach and pollution control measures are needed in the recycling sector. The industrial sector world widely is responsible for about one third of anthropogenic emissions of carbon dioxide, the most important greenhouse gas. Industry is also an important emitter of several other greenhouse gases. And many of industry's products emit greenhouse gases as well, either during use or after they become waste. Opportunities exist for substantial reducing industrial emissions through more efficient production and use of energy. Fuel substitutions, the use of alternative energy technologies, process modification, and by revising materials strategies to make use of less energy and greenhouse gas intensive materials. Industry has an additional role to play through the design of products that use less energy and materials and produce lower greenhouse gas emissions. All projects are required to have environmental impacts assessment conducted, covering all potential damage to the environment, mitigation and restoration, a reclamation plan including a resettlement programme for displaced residents, and the estimated implementation costs.

Table 1. Comparison between different energy consumed in Sudan (GWh)

Year	Electricity	Petroleum	Biomass	Renewable Technologies
1975	800	550	3000	50
1980	900	600	3500	60
1985	1200	650	4000	100
1990	1300	700	6000	120
1995	1400	800	7000	150
2000	1500	900	8000	200

Table 2. Classifications of data requirements

	Plant data	System data
Existing data	Size Life Cost (fixed and variable Operation and Maintenance) Forced outage Maintenance Efficiency Fuel Emissions	Peak load Load shape Capital costs Fuel costs Depreciation Rate of return Taxes
Future data	All of above, plus Capital costs Construction trajectory Date in service	System lead growth Fuel price growth Fuel import limits Inflation

Development in the environmental sense is a rather recent concern relating to the need to manage scarce natural resources in a prudent manner - because human welfare ultimately depends on ecological services. The environmental interpretation of sustainability focuses on the overall viability and health of ecological systems- defined in terms of a comprehensive, multiscale, dynamic, hierarchical measure of resilience, vigour and organisation. Natural resource degradation, pollution and loss of biodiversity are detrimental because they increase vulnerability, undermine system health, and reduce resilience. The environmental issues include:

- Global and transnational (climate change, and ozone layer depletion).
- Natural habitats (forests and other ecosystems).
- Land (agricultural zones).
- Water resources (river basin, aquifer, and water shed).
- Urban-industrial (metropolitan area, and air-shed).

Economic importance of environmental issue is increasing, and new technologies are expected to reduce pollution derived both from productive processes and products, with costs that are still unknown. This is due to market uncertainty, weak appropriability regime, lack of a dominant design, and difficulties in reconfiguring organisational routines. The degradation of the global environment is one of the most serious energy issues. Various options are proposed and investigated to mitigate climate change, acid rain or other environmental problems. Additionally, the following aspects play a fundamental role in developing environmental technologies, pointing out how technological trajectories depend both on exogenous market conditions and endogenous firm competencies:

(1) Regulations concerning introduction of Zero Emission Vehicles (ZEV), create market demand and business development for new technologies.
(2) Each stage of technology development requires alternative forms of division and coordination of innovative labour, upstream and downstream industries are involved in new forms of inter-firm relationships, causing a reconfiguration of product architectures and reducing effects of path dependency.
(3) Product differentiation increases firm capabilities to plan at the same time technology reduction and customer selection, while meeting requirements concerning network externalities.
(4) It is necessary to find and/or create alternative funding sources for each research, development and design stage of the new technologies.

Environmental sustainability depends on several factors, including:

- Climate change (magnitude and frequency of shocks).
- Systems vulnerability (extent of impact damage).
- System resilience (ability to recover from impacts).

The Sudan energy consumption is modest by international standards. The energy balance is dominated by biomass. Woody biomass makes up 71%, petroleum products 19%, non-woody biomass 8% and hydropower 2% [2]. The total Consumption of the country is as low as 6.3 million tons of oil equivalent (TOES) while the total primary supply is about 11 million TOES. This means that 43% of the energy is lost in converting wood to charcoal. Distances transported nearly always exceed 600 km. The major consumers are households (78% of the total energy). The largest consumers of firewood are brick markers. The efficiency of the traditional charcoal stoves is as low as 16 percent. The forest cover of Sudan receded from 25% in 1956 to 12% in 2001. Sudan is a member of the club of the 18 countries of "Least Forest Covered". Oil fields are in Addar Yale, Bashair and Heglig. Crude from the former two is transported by river barges and trucked respectively. The environmental impacts of the various processes involved have never been addressed. Further more some of the crude is wastefully used directly as a source of energy. The contribution of the two fields to the GNP has never been disclosed neither have their negative impacts on the environment. The oil is pumped from the various wells and fields through a system of 160 km of pipes the central facility at Heglig. The crude is de-watered at Heglig and then piped 1610 km to the Red Sea export terminal. The infrastructure of the operation includes an extensive network of raised roads and five field facilities. In addition to the central facility at Heglig, there is the airport, the contractors' camp, the newly-founded towns of Heglig and Keilak Al Kharassan, five pumping stations, a metering station at El Obeid, two refineries at El Obeid and Khartoum and the large refinery al El Gaili.

3. Sustainability

Absolute sustainability of electricity supply is a simple concept: no depletion of world resources and no ongoing accumulation of residues. Relative sustainability is a useful concept in comparing the sustainability of two or more generation technologies. Therefore, only renewables are absolutely sustainable, and nuclear is more sustainable than fossil. However, any discussion about sustainability must not neglect the ability or otherwise of the new technologies to support the satisfactory operation of the electricity supply infrastructure. The electricity supply system has been developed to have a high degree of resilience against the loss of transmission circuits and major generators, as well as unusually large and rapid load changes. It is unlikely that consumers would tolerate any reduction in the quality of the service, even if this were the result of the adoption of otherwise benign generation technologies. Renewables are generally weather-dependent and as such their likely output can be predicted but not controlled. The only control possible is to reduce the output below that available from the resource at any given time. Therefore, to safeguard system stability and security, renewables must be used in conjunction with other, controllable, generation and with large-scale energy storage. There is a substantial cost associated with this provision.

It is useful to codify all aspects of sustainability, thus ensuring that all factors are taken into account for each and every development proposal. Therefore, with the intention of promoting debate, a sustainability matrix is presented (Table 3). The following considerations are proposed:

(1) Long-term availability of the energy source or fuel.
(2) Price stability of energy source or fuel.
(3) Acceptability or otherwise of by-products of the generation process.
(4) Grid services, particularly controllability of real and reactive power output.
(5) Technological stability, likelihood of rapid technical obsolescence.
(6) Knowledge base of applying the technology.
(7) Life of the installation – a dam may last more than 100 years, but a gas turbine probably will not.
(8) Maintenance requirement of the plant.

4. Environmental Policies and Industrial Competitives

The industrial development strategy in Sudan gives priority to the rehabilitation of the major industrial areas with respect to improvement of infrastructure such as roads, water supply, power supply, sewer systems and other factors. This strategy also takes into consideration the importance of incorporating the environmental dimension into economic development plans. However, the relationship between environmental policies and industrial competitiveness has not been adequately examined. For the near future, the real issue concerns the effectiveness of environmental expenditures in terms of reduction of pollution emissions per unit of output. A number of issues relevant to this central concern are presented as follows:

Table 3. Sustainability matrixes

Power categories	1*	2*	3*	4*	5*	6*	7*	8*	9*	Index
Conventional coal fired stream plant	3	1	1	5	1	1	4	3	3	22
Oil fired stream plant	2	1	1	5	3	3	4	3	3	25
Combined cycle gas turbine	2	3	2	4	4	4	4	2	4	29
Micro combined heat and power	2	3	2	4	4	4	3	2	4	29
Nuclear	4	4	3	5	4	4	3	2	3	32
Hydropower	5	5	5	3	5	5	5	4	2	39
Tidal power	5	5	5	2	5	5	5	4	2	38
Onshore wind	5	5	5	2	5	5	4	4	3	38
Offshore wind	5	5	5	2	5	5	3	4	4	38
Land-fill gases	3	5	3	1	3	4	4	3	2	28
Municipal incineration	5	5	4	3	4	4	4	3	4	36
Biomass, field and forest crops plus waste straw	5	5	4	3	4	4	4	3	4	36
Import	1	1	5	1	5	5	5	5	5	33
Hydro pumped storage	-	-	5	5	5	5	5	5	2	32
Electrochemical storage	-	-	4	4	4	4	4	4	5	29
Diesel	2	1	1	1	4	5	3	4	4	25

1* fuel availability
2* price stability of fuel
3* by-product acceptability
4* grid services
5* technological obsolescence
6* knowledge base
7* life of the installation
8* maintenance requirement
9* infrastructure requirements

(1) Implementing Ecologically Sustainable Industrial Development Strategies

Agenda 21 for achieving sustainable development in the 21st century calls on governments to adopt National Strategies (NS) for sustainable development that ‘‘build on and harmonise the various sectoral, social and environmental policies that are operating in the country’’ [4]. NS Sudan focuses almost exclusively on development issues and does not integrate industrial and environmental concerns. It does not consider industrial specific environmental objectives or time frames for achieving them. Moreover, it does not specify how specific industrial subsectors and plants will meet environmental objectives. Finally, it is formulated with minimal involvement of industrial institutions and private-sector associations. To bring together industrial development and environmental objectives it is necessary to:

- Establish environmental goals and action plans for the industrial sector.
- Develop an appropriate mix of policy instruments that support the goals of those plans.
- Design appropriate monitoring and enforcement measurements to realise those goals.

(2) Applying Cleaner Production Processes and Techniques

Traditional approaches to pollution reduction have been based on the application of end of pipe technologies in order to meet discharge standards. However, the growing recognition that reduction at source is a potentially more cost effective method of abatement is resulting in replacing end of pipe technologies with cleaner production processes. Major constraints in adopting cleaner production methods relate to:

- Lack of awareness about the environmental and financial benefits of cleaner production activities.
- Lack of information about techniques and technologies.
- Inadequate financial resources to purchase imported technologies.

A coordinated effect by industry, government and international organisations can go a long way in overcoming these constraints. In this context key questions that need to be addressed are as follows:

(a) Need for local capacity building, information dissemination, training and education.
(b) Need for subsectoral demonstration projects.
(c) Need for increased cooperation with environmental market sectors in developed countries.
(d) Need for life cycle analysis and research on environmentally compatible products.

(3) Implementing Environmental Management Systems

Environmental management systems (EMSs) are necessary to enable plant to achieve and demonstrate sound environmental performance by controlling the environmental impact of their activities, products and services. The basic tools to ensure compliance with national and/or international requirements and continually improve its environmental performance include:

- Environmental auditing.
- Environmental reporting, and
- Environmental impact assessments.

In addition, the adoption of EMS may require extensive training of corporate staff. A practical and effective means of doing this is through the design and support of joint capacity strengthening programmes by industry association and bilateral and multilateral agencies.

(4) Managing and Conserving Water Resources

It is estimated that by year 2025, there will be a global crisis in water resources. Accelerated growth of industry will lead to increase in industrial water use. Moreover, major industrial water pollutant load is expected to increase considerably in the near future. Therefore, to better manage water resources by industry, there is a real need for integrating demand trend and use patterns. The main elements of an industrial management strategy can be identified as follows:

- Analytical services.
- Promotional services.
- Services for the development of industry and water supply infrastructure.

(5) Using Market Based Instruments (MBIs) to Internalise Environmental Costs

As complements to command and control measures for resource conservation and pollution prevention in industry. MBIs represent a useful and efficient cost effective policy measures that internalise environmental costs. A plant's decision to invest in clean production depends primarily on the following factors:

(a) Relative costs of pollution control in overall production costs.
(b) Price elasticities of supply and demand for intermediary and final goods, and
(c) Competitive position of plant in a particular industrial sector.

(6) Counteracting Threats from Eco-labelling Requirements

The increasing export orientation of production makes it necessary to maintain competitive position in world markets. The emergence of a wide variety of eco-labelling requirements and lack of timely information on multitude of scheme may adversely affect certain export sectors. Needed initiatives to counteracting perceived threats could be presented as follows:

- Information dissemination.
- Life cycle analysis.
- Establishing certification centres.
- Infrastructure support.

The National Council for Environment has been established to develop a legal framework and policy of the Federal Environmental Protection since 1990. There is a legal framework in the country covering environmental impact studies for water resource development projects, and local people are informed about plans for new or refurbishment schemes through newspapers, radios, and television.

(7) Implementing the United Nations (UN) Framework Convention on Climate Change

The UN climate change convention entered into force on 21st March 1994. The convention objective is the stabilisation of greenhouse gas concentration in the atmosphere at safe levels. For industry, responding to this convention will undoubtedly be a major challenge. Industry will be directly affected. Sudan as party to this convention is obliged to take a number of actions and cooperates effectively in order to meet this challenge. Sudan has to contribute to the common goal of reducing greenhouse gases emissions by taking precautionary measures to mitigate causes and anticipate impacts of climate change. However, there may not be adequate means to do so, and Sudan will therefore require international assistance. The main requirements are:

- Access to best energy-efficient technologies available on the world market, where such technologies are relevant to the natural resources endowments, the industrial requirements and are cost effective.
- Building an energy-efficient capital stock by accelerating the development of low energy intensity processes and equipment.
- Strengthening national capabilities for energy-efficient design and manufacturing.

Areas where technical expertise to implement the convention is necessary include:

- Preparing national communications on greenhouse gas emissions. The communications are supported to contain an assessment of the magnitudes and sources of greenhouse gases as well as identification of reduction methods.
- Supporting technology transfer for improvement in the efficiency of fuel based power generation.
- Promotion technology transfer for the use of renewable sources of energy such as biomass, wind, solar, hydro, etc.
- Developing and implementing technology transfer for energy efficiency programmes in industry, in complementarities with cleaner production/pollution prevention measures.
- Analysing the impact of climate change response measures on the economic and industrial development of the country, with the view to identifying economically viable technology options for reducing greenhouse gas emissions from the production and consumption of energy.

(8) Addressing Concerns of Small and Medium Scale Industry (SMI)

Small and medium scale enterprises not only contribute to productivity growth and employment but are also important as collective sources of localised pollution loading such as organic wastes in water effluent, as well as hazardous wastes, heavy metal sludge, solvents, waste oils, acidic and alkaline wastes, photo wastes, etc. Often, these wastes are disposed of in unsafe manure and are extremely difficult to monitor. The cost of control in relation to output is too high, so even a modest increase in the costs (of environmental regulations) may threaten prevention and control may be well known and easily available, there is no guarantee that they will be adopted. Moreover, even when policy measures are in place, their enforcement and monitoring is a real problem for SMI sector on account of their large numbers and diversity. It is clear that environment problems of SMIs require special attention and special measures to address their particular problems.

5. Petroleum Industry Pollution and Greenhouse Gases Emissions in Sudan

The activities of oil exploration in Sudan began in late 1950s in the coastal areas of Red Sea. The results of exploration indicated that there is considerable amount of natural and liquefied gases in Suwakin and Bashair, and the quantities were estimated between 45-326 x 10^9 cubic meters. According to the increasing oil industry activities in Sudan such as production, refining and export/consumption, and if we consider the entire fuel cycle, namely: exploration, extraction, preparation/transformation, transportation, storage, pollution, including the increase in greenhouse gases, as result of petroleum industry will be very significant in the forthcoming future. In the year 1997 about 2 x 10^9 tonnes of petroleum products were burnt in Sudan. This amount will be doubled in the year 2010. More information and data concerning the area of greenhouse gases recording in Sudan are available in the First National Communication 2003 and Second National Communication under preparation by the Higher Council of Environment and Natural Reseources.

6. Privatisation and Price Liberalisation in Energy Source Supplies

The privatisation, and price liberalisation in energy fields has been secured to some extent (but not fully). Availability and adequate energy supplies to the major productive sectors are still needed. The result is that, the present situation of energy supplies is for better than ten years ago. The investment law has also encourage of the participation of the investors from the national level as well as from the international friendly and sisters' countries to invest in energy supply system such as:

- Petroleum products (import in particular) in the northern states.
- Electricity generation (in some states) using large diesel engine driven generators.

The readily implementation of electricity price liberalisation has relieved to some extent the National Electricity Corporation (NEC) from the heavy dependency of government subsidies, and a noticeable improved of the NEC management, and electricity supplies are achieved. In a country with a dense population, there are extreme pressures on energy supply system, which can stunt the country's economic growth. However, Sudan has recognised the potential to alleviate some of these problems by promoting hydro energy and utilising its vast and diverse climate, landscape and resources and by enhancing these for energy production.

7. Synthesis of the Renewable Energy

Although the overall impact of renewables has been necessarily low, the experience has clearly demonstrated their potential as sustainable energy alternatives. There has been substantial learning in disseminating and managing various technologies on account of:

- Scale: with increasing numbers, teething problems have been overcome and better knowledge has been gained in different aspects related to planning, implementation, operation and maintenance.
- Indigenisation: through joint ventures with international industry, the technology transfer process has been facilitated, helping in developing local production capacities.
- Infrastructure: a strong infrastructure has been created over the years to provide the technical, operational and managerial support to intervention programmes. This includes research institutions, training agencies, NGOs, financial intermediaries, etc.
- Diverse strategies: though the whole renewable energy programme started with the same technology push approach, diversification occurred over a period of time in terms of strategies and to promote different technologies according to market conditions (Table 4).

The use of geothermal energy in Sudan has the following special characteristics:
Disadvantages:

- Geothermal energy is a low-temperature resource tied to specific locations and can be used only in specific types of system.
- Minerals in some thermal waters can produce scaling problems that increase costs.

Advantages:

- Sudan has a big geothermal energy potential.
- Geothermal energy can be environmentally sound if wastewater or brine is properly disposed of (e.g., by reinjection) and there is no air pollution.
- The dependency on imported energy sources could be decreased.
- The number of jobs can be increased.
- Competitiveness of geothermal energy should increase owing to projected increases in domestic prices of competing fossil fuels.

Not only does cost of the fuel drive the researcher to explore an alternative source of energy, but ozone layer problems is other factor that should be considered in searching for cheap and clean energy sources (Table 5). Solar energy is an unlimited, cheap and clean source of energy, which has been utilised to replace conventional energy, but there still is a need to develop a technology to utilise the solar energy. Nowadays, solar energy is widely used in heating water, dehumidifying air and generating electrical energy for many domestic, agricultural and industrial applications. For these applications, the most important piece of the system equipment is the solar collector in which solar energy is converted into heat or electrical power. Although there are several designs for solar energy collecting devices, flat-plate collectors are the most common and popular type. They have the advantages of being simple to build, employing locally available materials, they are easy to operate and maintain, they have the ability to function even during cloudy or hazy days with diffuse solar radiation only, and they are capable of integration into a roof or a wall structure.

Fresh water supply is one of the most limiting conditions for the populations of arid regions. Following methods can solve the problem of providing arid areas with fresh water:

(a) Transportation of water from other locations.
(b) Desalination of saline water (ground and underground).
(c) Extraction of water from atmospheric air.

Transportation of water through these regions is usually very expensive, and desalination depends on the presence of saline water resources, which are usually rare in arid regions. The extraction of water from atmospheric air can be accomplished by two methods: cooling moist air, and absorbing water vapour from moist air using a solid or a liquid desiccant, with subsequent recovery for the extracted water by heating the desiccant and condensing the evaporated water. The significant increase in recent years of the number of rural electrification systems, using photovoltaic technology for illumination or water pumping in order to reduce the burden of capital costs per unit of generated power.

Table 4. Diversity of promotion strategies

Technology	Strategy	Key change agents	Mechanisms
Power generation			
Wind	Government-enabled market pull	Private sector State utilities Manufacturers	Fiscal incentives Multilateral finance
Small-hydro	Government-enabled market pull	Private sector State utilities Manufacturers	Fiscal incentives Multilateral finance
Biomass cogeneration	Government-enabled market pull	Private sector	Fiscal incentives
Solar PV	Combination of demonstration and pull	Government NGOs Intermediaries Manufacturers	Low interest finance Flexible loan serving Maintenance by Intermediaries
Thermal energy			
Biogas	Technology push	Government NGOs Turnkey workers	Cash subsidy Training support
Improved stoves	Technology push	Government NGOs Self-employed workers	Cash subsidy Training support
Solar cookers	Push in the beginning Currently market pull	Government Manufacturers	Mini-subsidy Segmentation with focus On small towns and cities
Solar water heaters	Push in the beginning Currently market pull	Government Manufacturers	Segmentation with focus on industrial systems

The power generation and supply for isolated areas can be effectively achieved using wind energy as the principle energy source. The savings are not only economic but also environmentally sound. However, the transient nature of the energy supply that is delivered to the power system from wind-turbine poses significant challenges to the design and control of the system. The efficiency of such systems can only improve with more effective control and a better understanding of the characteristics of these types of systems.

Considering energy crises and pollution problems today, investigations have been concentrated on decreasing fuel consumption by using alternative fuels and lowering the concentration of toxic components in combustion products. Hydrogen is considered as a renewable clean energy source. Two methods using hydrogen in spark ignition engines were investigated in Sudan. The first method is to mix hydrogen with gasoline. The second concerns using hydrogen as a standard fuel. The advantage of the hydrogen-supplemented fuel is that it requires a smaller quantity of hydrogen, which considerably reduces the problems connected with hydrogen storage in the automobile. A hydrogen-gasoline fuelled engine generally develops lower maximum power and higher NO_x emission compared to an equivalent gasoline engine. Hydrogen can be used as a supplementary fuel in modern spark ignition engines without major changes and it can help save a considerable part of the available oil and save the environment from toxic pollutants.

Depletion of fossil fuels and fluctuating prices has rekinded an interest in the development of renewable fuels/energy sources such as biomass. There is a critical need for development of alternative, appropriate, affordable methods of cooking for use in Sudan. Population pressures on remaining forest resources have resulted in shortages of fuel available for household use and yielded adverse environmental effects. Solar cooker use has the potential to alleviate both problems by reducing dependence on wood as a cooking fuel. However, to gain acceptance and motivate use, there is a need to develop many different solar cooker designs. Each design needs to be suited to specific climates, customers and economic factors. The development requires a good fundamental understanding of the relationship between key design variables and performance. The most commonly used type of solar cooker is the box type. Box type cookers depend on heat retention. They are slow to heat up, but work well even where there is diffuse radiation, convective heat loss caused by wind, intermittent cloud cover, and low ambient temperatures. Despite the variation in approach flat plate, concentrating, and box type, and large variety of inventions and products, the total potential design space is largely unexplored. Systematic examination of the design space requires either extensive testing or models with broad predictive capacity.

The availability of methane and/or natural gas has lead to a worldwide spread of internal combustion engines running on the dual fuel concept. Gaseous fuels also promise to be suitable for higher compression engine since it is known that they resist knock more than conventional liquid fuels, as well as producing less polluting exhaust gases, if appropriate conditions are satisfied for its mixing and combustion. Therefore, it is more economical and of environmental advantage to use natural gases in diesel engines, which use the dual fuel concept. Energy efficiency brings health, productivity, safety, comfort and savings to the homeowner, as well as local and global environmental benefits. The use of hydro energy resources could play an important role in this context, especially with regard to responsible and sustainable development. It represents an excellent opportunity to offer a higher standard of living to the local people and will save local and regional resources. Implementation of hydro energy offers an opportunity for economic improvement by creating a market for producing companies as well as maintenance and repair services.

8. Climate Change, Global Warming and The Enhanced Greenhouse Effect

Changes in the global climate (mean temperature, precipitation, etc.) could also threaten stability of a range of critical interlinked physical, ecological and social systems and subsystems [5]. An important social principle is that the climate change should not be allowed to worsen existing inequities- although climate change policy cannot be expected to address all prevailing equity issues. Some special aspects include:

- The establishment of an equitable and participative global framework for making and implementing collective decisions about climate changes.
- Reducing the potential for social disruption and conflicts arising from climate change impacts.
- Protection of threatened cultures and preservation of cultural diversity.

Table 5. Correlation of solar radiation with other weather parameters in Sudan (Yearly averages)

Station	Mean temp. (°C)	Sunshine duration (h)	Solar radiation (MJm^{-2} day^{-1})	Wind velocity (ms^{-1})	Relative humidity (%)
Port Sudan	28.4	9.0	20.87	5.1	65
Shambat	29.7	9.9	22.82	4.5	31
Wad Medani	28.4	9.8	22.84	4.5	40
El Fasher	25.8	9.6	22.80	3.4	33
Abu Na'ama	28.2	8.8	21.90	3.1	46
Ghazala Gawazat	27.2	9.3	21.72	3.0	43
Malakal	27.9	7.8	19.90	2.8	54
Juba	27.6	7.8	19.59	1.5	66
Dongola	27.2	10.5	24.06	4.6	27
Toker	28.8	7.3	17.60	4.1	53
Hudeiba	29.3	10.0	22.37	4.0	25
Aroma	29.1	9.6	21.40	4.2	37
El Showak	26.3	9.7	22.90	4.1	39
Zalingei	24.5	8.8	22.98	2.7	39
Babanusa	28.2	8.9	21.73	2.8	48
Kadugli	27.5	8.5	21.30	2.7	48

Climate change is one of the most serious concerns of our society. The gases causing the greenhouse effect are carbon dioxide, methane, nitrous oxide, chlorofluorocarbons, tropospheric ozone, and stratospheric water vapour. Carbon dioxide is an important greenhouse gas and a major agent of climate change and the most significant greenhouse gas contributing about half of the total greenhouse effect [6]. The recent concentration of atmospheric CO_2 is 25% higher than the pre-industrial level [7]. This increase is primarily due to fossil fuel combustion and deforestation. Climate change might pose a serious threat to the ecological and socio-economic systems, unless measures are investigated to mitigate the rising accumulation of atmospheric CO_2. Green-space in urban ecosystems can reduce atmospheric CO_2 levels in three ways. Here, green-space is defined as soil surface area capable of supporting vegetation and the vegetation being supported. First, urban trees and shrubs directly sequester and accumulate atmospheric CO_2 in the process of their growth through photosynthesis. Second, urban vegetation decreases building cooling demand by shading and evapotranspiration, and heating demand by wind speed reduction, thereby reducing CO_2 emissions associated with fossil fuel use. Third, urban soils store organic CO_2 from litter-fall, until it is returned to the atmosphere by decomposition. On the other hand, buildings, factories, and automobiles in urban landscapes release CO_2 through fossil fuel consumption.

Industry's use of fossil fuels has been blamed for our warming climate, when coal, gas and oil are burnt, they release harmful gases, which trap heat in the atmosphere and cause global warming. However, there has been an ongoing debate on this subject, as scientists have struggled to distinguish between changes, which are human induced, and those, which could be put down to natural climate variability. Industrialised countries have the highest emission levels, and must shoulder the greatest responsibility for global warming. But action must also be taken by developing countries to avoid future increases in emission levels as their economies develop and population grows. Human activities that emit carbon dioxide (CO_2), the most significant contributor to potential climate change, occur primarily from fossil fuel production. Consequently, efforts to control CO_2 emissions could have serious, negative consequences for economic growth, employment, investment, trade and the standard living of individuals everywhere. Scientifically, it is difficult to predict the relation between global temperature and greenhouse gas concentrations. The climate system contains many processes that will change if warming occurs. Critical processes include heat transfer by winds and currents, the hydrological cycle involving evaporation, precipitation, runoff and groundwater, and the formation of clouds, snow, and ice, all of which display enormous natural variability. The equipment and infrastructure for energy supply and use are designed with long lifetimes, and the premature turnover of capital stock involves significant costs. Economic benefits occur if capital stock is replaced with more efficient equipment in step with its normal replacement cycle, and if opportunities to reduce future emissions are taken in the world they should be less costly. Such flexible approaches would allow society to take account of evolving scientific and technological knowledge, and to gain experience in designing policies to address climate change. Focusing only on CO_2 emissions neglects the full richness of the environmental data, but has been done for several reasons:

- CO_2 is an important global environmental indicator.
- CO_2 emissions can show very significant increases, as Sudan currently produces very little CO_2 from the electricity sector.
- CO_2 and other air pollutants are linked together to new generation capacity, So CO_2 serves as a surrogate for most other emissions.

The environmental characteristics of products have become increasingly important to consumers [8, 9]. Firms have responded by placing eco-labels on products that highlight the item's environmental attributes and by introducing new, or redesigned green products [10]. Governments and non-governmental organisations have also responded by organising, implementing, and verifying eco-labelling programmes that cover thousands of products, which international efforts to standardise environmental labelling schemes have also emerged. From a policy perspective, one aim of eco-labels is to educate consumers about the environmental impacts of the product's manufacture, use, and disposal, thereby leading to a change in purchasing behaviour and ultimately, to a reduction in negative impacts. Further, eco-labelling policies may promote environmental objectives without production site command and control methods and are seen a way of meeting global environmental objectives while complying with international trade agreements. Recent techniques for economically valuing environmental impacts:

- Effect on production.
- Effect on health.
- Defensive or preventive costs.
- Replacement cost and shadow projects.
- Travel cost.
- Property value.
- Wage differences (the wage differential method attempts to relate changes in the wage rate to environmental conditions, after accounting for the effects of all factors other than environment (e.g., age, skill level, job responsibility, etc.) that might influence wages).

9. Major Energy Consuming Sectors

Sudan is still considered as one of the 25 least developed African countries. Agriculture is the backbone of economic and social development in Sudan. About 80% of the population depends on agriculture, and all other sectors are largely dependent on it. Agriculture contributes to about 41% of the gross national product (GNP) and 95% of all earnings. Agriculture determines for the last 30 years the degree of performance growth of the national economy.

9.1. Agricultural Sector

During the last decades agriculture contributed by about 41% to the Sudan GNP. This share remained stable until 1984-1985 when Sudan was seriously hit by drought and desertification, which led to food shortages, deforestation, and also, by socio-economic effects caused the imposed civil war. The result dropped the agriculture share to about 37%. Recent development due to rehabilitation and improvement in the agricultural sector in 1994 has raised the share to 41%. This share was reflected in providing raw materials to local industries and an increased export earning besides raising percentage of employment among population. This sector consumes 2.5% of the total energy consumption (28% from electricity, 14.8% from fossil fuels, and the rest from biomass fuels) [11].

9.2. Industrial Sector

The industrial sector is mainly suffering from power shortages, which is the prime mover to the large, medium and small industries. The industrial sector is consuming 5.7% of the total energy consumption, distributed as follows: 55% from petroleum products, 13% from biomass and 32% from electricity [12].

9.3. Domestic Use

Household is the major energy consumer. It consumes 92% of the total biomass consumption in the form of firewood and charcoal. From electricity this sector consumes 60% of the total consumption, and 5.5% of petroleum products.

9.4. Transport Sector

The transportation sector (railways, ships, boats, etc.) was not being efficient for the last two decades because of the serious damage that happened to its infrastructure (roads, workshops, and maintenance centres, etc.). It consumes 10% of the total energy consumption and utilises 60% of the total petroleum products supplied.

Sudan is an energy importing country and the energy requirements has been supplied through imports that have caused financial problems. Because of the economical problems in Sudan today, the Sudanese energy policy should be concentrated on assurance of energy supply, reliability, domestic sufficiency, in time, in economic terms, and renewability. Therefore as a renewable energy source, biomass (especially fuelwood) seems interesting because its share of the total energy production at 87% is high and the techniques for converting it to useful energy are easy. On the other hand, biomass may, however, see greatly expanded use in response to the environmental problems caused by fossil fuel use in the country. Biomass has been proposed to have a central role to play in future, more sustainable energy scenarios. For this to become a reality several real problems need to be overcome. In Sudan as in other developing countries modernisation of biomass energy provision is an urgent necessity for the sake of human health, protection of the environment, and climate change abatement. Given sufficient recognition, resources and research biomass could become the environmentally friendly fuel of the future.

10. Mitigation Measures

Mitigation measures that could be under taken to influence the effect of oil industry and use that may contribute in decreasing greenhouse gases (GHGs) emissions and decelerate the threat of global climate change may include the following:

- Controlling GHGs emissions by improving the efficiency of energy use, changing equipment and operating procedures.
- Controlling GHGs emission detection techniques in oil production, transportation and refining processes in Sudan.
- More efficient use of energy-intensive materials and changes in consumption patterns.
- A shift to low carbon fuels, especially in designing new refineries.
- The development of alternative energy sources (e.g., biomass, solar, wind, hydro-electrical and cogeneration).
- The development of effective environment standards, policies, laws and regulations particularly in the field of oil industry.
- Activating and supporting environmental and pollution control activities within the Ministry of Energy and Mining (MEM) to effectively cope with the evolving oil industry in Sudan.

11. Policy Development

The non-technical issues, which have recently gained attention include:

- Environmental and ecological factors, e.g., carbon sequestration, reforestation and revegetation.
- Biomass as CO_2 neutral replacement for fossil fuels.
- Greater recognition of the importance of renewable energy, particularly modern biomass energy carriers, at the policy and planning levels.
- Greater recognition of the difficulties of gathering good and reliable renewable energy data, and efforts to improve it.
- Studies on the detrimental health efforts of renewable energy particularly from traditional energy users.
- Greater awareness of the need to internalise the externality cost of conventional energy carriers to place them on more equal terms with alternative energy sources.

12. The Future

(1) In the most of the developing countries, the governments acknowledge that, renewable energy can resolve many pressing problems. Yet, the matter stops at this level ''Acknowledgement''. Much more is needed, like laws regulating and

encouraging business, tax concessions, both to investors and customers, and most of all, a sustained, coordinated and well-planned official publicity campaign to enlight, inform and educate the public at a large.

(2) To avoid the problems of fuel altogether (uncertain availability and skyrocketing prices), and minimise spare-parts, solar and wind pumps are proposed to replace diesel engines in the predominant irrigation areas.

(3) Local manufacture, whenever possible, is to be emphasised to avail renewable energy devices since limited funds are the main constraints in commercialisation and dissemination of the technology. Low cost devices as well as reliable devices have to be provided.

(4) Embarking on conservation energy and reduction of pollution of environment to be undertaken without delay:

- To save on fossil fuel for premium users/export.
- To accelerate development of new and/or remote lands otherwise deprived of conventional energy sources.
- As a preventive measure against shortage of future energy supply against prospective national energy demand.

(5) Launching of public awareness campaigns among investor's particularly small-scale entrepreneurs and end users of renewable energy technologies to highlight the importance and benefits of renewables.

(6) To direct Sudan resources away from feeding wars and the arms industry towards real development, this will serve the noble ends of peace and progress.

(7) The energy crisis is a national issue and not only a concern of the energy sector, and the country has to learn to live with the crisis for a long period, and develop policies, institutions and manpower for longer term, more effective solutions.

(8) To invest in research and development through the existing specialised bodies, e.g., Energy Research Institute (ERI).

(9) To encourage co-operation between nations, a fact this will be much easier in this era of information and the communications revolution.

(10)Government should give incentives to encourage the household sector to use renewable energy technologies instead of conventional energy.

(11)Promotion research and development, demonstration and adaptation of renewable energy resources (solar, wind, biomass, and mini-hydro, etc.) amongst national, regional, and international organisations which seek clean, safe, and abundant energy sources.

(12)Execute joint investments between the private sector and the financing entities to disseminate the renewables with technical support from the research and development entities.

(13)Promotion the general acceptance of renewable energy strategies by supporting comprehensive economic energy analysis taking account of environmental benefit.

(14)Availing of training opportunities to personnel at different levels in donor countries and other developing countries to make use of their wide experience in application and commercialisation of renewable energy technologies.

(15)To encourage the private sector to assemble, install, repair and manufacture renewable energy devices via investment encouragement, more flexible licensing procedures.

(16)In order to achieve the major changes needed to alleviate the environmental impacts of the building sector, it is necessary to change and develop both the processes in the industry itself, and to build a favourable framework to overcome the present economic, regulatory and institutional barriers.

13. Conclusions

Air pollution from motor vehicles, electricity generation plants, industry, and other sources natural and man-made, can harm human health, injure crops and forests, damage building materials, and impair visibility. Public awareness and concern about the problems associated with reduced air quality have increased in recent years. Nevertheless, there still remains considerable uncertainty about both the severity and the valuation of these impacts. The valuation of environmental damages can play an important role in establishing environmental policy and regulatory standards, and can provide guidance in targeting mitigation efforts. In order to achieve environmental objectives at least cost, policy-makers and managers need to balance the relevant social costs and benefits.

The mitigation strategy of the country should be based primarily ongoing governmental programmes, which have originally been launched for other purposes, but may contribute to a relevant reduction of greenhouse gas emissions (energy-saving and afforestation programmes). Therefore, the main fields of emission mitigation will be the energy and the forestry sectors. The main reasons are as follows:

The overall energy efficiency is far lower than that of the industrialised market economies. The efficiency on both the demands and the supply side have to be increased even in the short run. As far as the supply side is concerned, restructuring of the power plant system is unavoidable, since many plants are old.

Traditionally, the forestry sector is highly developed and there is enough land even for a larger afforestation programme.

References

[1] Commission of the European Communities. (2000). Towards a European strategy for the security of energy supply. *Green Paper*, COM (2000), 769, Brussels, 29 November 2000.

[2] Gheorghe A., and Nicolet-Monnier M. (1995). *Integrated regional risk assessment.* Vols I and II, Kluwer academic, Dordrecht.

[3] DECADES Project on comparative assessment of various energy systems. (1997). IAEA, Vienna.

[4] UNIDO. (1997). Changing courses sustainable industrial development, as a response to Agenda 21, Vienna.

[5] IPCC. (1996). Climate change 1995: impacts, adaptations and mitigation of climate change. Watson, R.T. et al., (Editors). Cambridge University Press, London.
[6] Rodhe, H. (1990). A comparison of the contributions of various gases to the greenhouse effect. *Science, 248,* 1217-1219.
[7] Post, W. H., Peng, T. H., Emanuel, W. R., King, A. W., Dale, V. H., and DeAngelis D. L. (1990). The global carbon cycle. *American Scientist 78,* 310-326.
[8] Loro, L. (1993). Green marketing comes to computers. *Advertising Age*, September 29.
[9] U.S.A Environmental Protection Agency. (1994). Determinants of effectiveness for environmental certification and labelling programmes.
[10] U.S.A Environmental Protection Agency. (1993). Status report on the use of environmental labels worldwide.
[11] Omer, A. M. (2008). Green energies and the environment. *Renewable and Sustainable Energy Reviews* 12: 1789-1821.
[12] Omer, A. M. (2008). *Energy demand for heating and cooling equipment systems and technology advancements.* In: Natural Resources: Economics, Management and Policy, p.131-165.

In: Energy Costs, International Developments...
Ed: Leszek Kowalczyk and Jakub Piotrowski

ISBN: 978-1-60741-700-2

Chapter 5

ALTERNATIVE ENERGY FOR FUTURE

Abdeen Mustafa Omer
17 Juniper Court, Forest Road West, Nottingham NG7 4EU, UK

ABSTRACT

Ground source heat pumps (GSHPs) provide an effective and clean way of heating buildings worldwide. They make use of renewable energy stored in the ground, providing one of the most energy-efficient ways of heating buildings. They are suitable for a wide variety of building types and are particularly appropriate for low environmental impact projects. They do not require hot rocks (geothermal energy) and can be installed in most of the world, using a borehole or shallow trenches or, less commonly, by extracting heat from a pond or lake. Heat collecting pipes in a closed loop, containing water (with a little antifreeze) are used to extract this stored energy, which can then be used to provide space heating and domestic hot water. In some applications, the pump can be reversed in summer to provide an element of cooling. The only energy used by the GSHP systems is electricity to power the pumps. Typically, a GSHP will deliver three or four times as much thermal energy (heat) as is used in electrical energy to drive the system. For a particularly environmental solution, green electricity can be purchased. The GSHP systems have been widely used in other parts of the world, including North America and Europe, for many years. Typically they cost more to install than conventional systems; however, they have very low maintenance costs and can be expected to provide reliable and environmentally friendly heating for in excess of 20 years. The GSHPs work best with heating systems, which are optimised to run at a lower water temperature than is commonly used in the UK boiler and radiator systems. As such, they make an ideal partner for underfloor heating systems. This article reviews seasonal thermal energy storage underground heat, and the potential of using heat pipe technology into the GSHP system. This research also covers soil thermal properties, estimated ground temperatures and methods of system control. It would appear to be a logical unit to rebate for a winter or summer peaking utility. The air heat pumps cannot match the lower cost of operating typical electric/gas equipment. The GSHP reduces demand and energy. Energy savings are sufficient enough to warrant added investments in the $600 to $900/ton range, especially if costs can be assumed in a 15 or 30-year mortgage.

Keywords: GSHPs, potential, and applications, technology, and environment.

1. Introduction

With improvement of the people's living standards and development of the economies, heat pumps have become widely used for air conditioning. The driver to this was that environmental problems associated with the use of refrigeration equipment, the ozone layer depletion and global warming are increasingly becoming the main concerns in developed and developing countries alike. With development and enlargement of the cities in these cold regions, the conventional heating methods can severely pollute the environment. In order to clean the cities, the governments drew many measures to restrict citizen heating by burning coal and oil and to encourage them to use electric or gas-burning heating. New approaches are being studied and solar assisted, reversible absorption heat pump for small power applications using water-ammonia is under development [1].

An air-source heat pump is convenient to use and so it is a better method for electric heating. The ambient temperature in winter is comparatively high in most regions, so heat pump with high efficiency can satisfy their heating requirement. In fact, a conventional heat pump is unable to meet the heating requirement in severely cold regions, because it's heating capacity decreases rapidly when ambient temperature is below $-10^{o}C$. According to the weather data, the air-source heat pump for heating in cold regions must operate for long times with high efficiency and reliability when ambient temperature is as low as $-15^{o}C$. The conventional heat pump used in warm regions cannot operate steadily in such cold regions to satisfy the heating requirement in winter. Accordingly, many researches and developments have been conducted to enable heat pump to operate steadily with high efficiency and reliability in low temperature environment [2]. For example, the burner of a room air conditioner, which uses kerosene, was developed to improve the performance in low outside temperature [3]. Similarly, the packaged heat pump with variable frequency scroll compressor was developed to realise high temperature air supply and high capacity even under the low ambient temperature of -10 to $-20^{o}C$ [4]. Such a heat pump system can be conveniently used for heating in cold regions. The importance of targeting the usage for the low capacity range is clear if one has in mind that the air conditioning units below 10 kW cooling account for more than 90% of the total number of units installed in the EU [5].

2. Energy Extraction from Cold Water

Climate change is a real threat to the planet's future, and a major cause for it is the use of fossil fuels to power homes and businesses. Renewable energy, combined with energy efficiency, offers a viable and potent solution to countering the effects of global warming. By installing any of the available renewable energy technologies, one will be making a major personal contribution to the well-being of future generations and could also benefit from lower fuel bills.

In heating applications, heat pumps save energy by extracting heat from a natural or waste source, using a mechanism similar to that found in a refrigerator. They can be used for any normal heating needs. However, this technology is not new. Several heat pumps were installed in the 1950's in a bid to save energy and fuel costs. One of the most famous of these was used to heat the Royal Festival Hall in London by extracting heat from the River Thames [6].

Humans' natural sense of heat is based more on instinct than science. Humans are warm-blooded and judge "heat" by comparing it to touch. Since the body temperature needs to be maintained to within a few degrees Celsius, the natural senses have evolved to make extremes of temperature uncomfortable. Hence, a hot summer's day feels many times "hotter" than the freezing mid-winter. But in reality the earth's surface does not vary in "heat energy" as much as one might imagine. Scientifically speaking, there is only 11% less energy in cold river water at 5°C (40°F) compared to hot bath water at 40°C (105°F).

The most familiar form of heat pump is the domestic refrigerator. The heat is extracted from the cabinet to keep food fresh and the extracted heat is expelled through the radiator grill at the back of the unit. In this case, the heat is merely a waste product. The heat pump, utilises this heat, and put the "cold part" outside. To make this more understandable, imagine that the "ice box" of a refrigerator is immersed in a small garden stream and the hot grid from the back is placed inside a house. The "ice box" will attempt to freeze the stream and, if the stream stopped from flowing, freezing of the water would naturally occur. But the passing water will constantly warm up the very cold "ice box". Hence, the temperature of the stream will be reduced immeasurably. So, heat is extracted from the stream, which ends up as heat in the radiator grill, available to warm the house. In every case, the useful heat output will be greater than the energy required to drive the heat pump itself. Therefore, the heat was extracted from the stream for "free". Another way to think of it is as follows. If an electric kettle element was immersed in the stream, then water will rapidly absorb any warmth from the element. This would be a one-way loss of energy to the stream. If, conversely, the element is colder than the stream, then the stream will warm it up. Therefore, the surrounding will absorb, hence gaining energy.

There are various types of heat pump with many different uses. Nearly all heat pumps use electricity as the form of energy input. Types using gas engines are almost unheard-of in the UK. These can save significant amounts of energy, and are a thing for the future. The heat pump usually delivers heat in the form of hot water. To maintain high-energy efficiency, the system should be designed so that the water temperature is not too high. For this reason, radiators with larger than average surface area (or more of them) should be used. Water temperatures within the pipes of an underfloor heating system can be as low as 35 Celsius (95°F); this gives a very high efficiency. Such systems are very comfortable, and are especially good when used in well-insulated houses.

Swimming pool applications are very energy-efficient and fairly common. These are mostly air-source, taking heat from the air. An air source system, however, will be less effective in winter since the air temperature fluctuates and becomes very cold. For a year-round heating, a river, stream or spring is much more stable and better heat source, offering higher efficiency. The water source should ideally be close to the property. The water flow required is less than one litre per second for a 10 Watt output heat pump, so the smallest of streams can be utilised. Springs and boreholes can often deliver water at a steady 10 degrees Celsius (50°F) throughout the winter, making them excellent heat sources. Ground pipes, buried either vertically or horizontally, are good heat sources. This type is often referred to as geothermal. This system is particularly suited if cooling is also required since in cooling mode (air-conditioning), the energy efficiency is significantly better as compared to a conventional air cooled systems (Figure 1).

3. ENERGY SOURCE: THE GROUNDS

In the autumn of 2000, The British Engineering Council awarded an Environmental Engineering award to the GSHP project at Commerce Way, Surrey. This largest UK ground source project is a speculative built industrial building of about 3000 m^2 with both offices and warehouse facilities. It is known that during the normal life span of a building the surplus of heat would lead to higher ground temperatures. This leads to less efficient heat pump operation and may result in insufficient capacity during cooling and peak demands. As a solution, the project developed a hybrid system incorporating a dry-cooler [6]. The principle idea was to use the dry-cooler to store cold in the wellfield during early spring, when the required summer peak load cool can be generated very efficiently and cheaply. A geothermal energy system uses the ground as a heat-source or heat sink, depending on whether the systems were used in a heating or cooling mode. The ground is principally suited for low temperature energy exchange. The usual operating temperature bandwidth is between –5°C and 40°C (not taking into account high temperature energy stores). Different systems for exchanging energy with the ground are currently in use, such as direct use of groundwater, closed-loop ground heat exchangers or direct expansion [6].

There are various sources of energy in the form of stored solar energy, which can provide heat even if they have a low temperature. But, how can a few degrees above zero give enough energy to heat radiators and hot water? During the summer, solar heat is stored in the surface layer of the soil. Also, down in the bedrock there is a source of heat that stays at practically the same temperature all year round. With the help of the GSHPs one can use either or both of these energy sources, as explained below. The most common type of heat pump-air-source heat pump-uses outside air as the heat source during the heating season and the heat sink during the air-conditioning season.

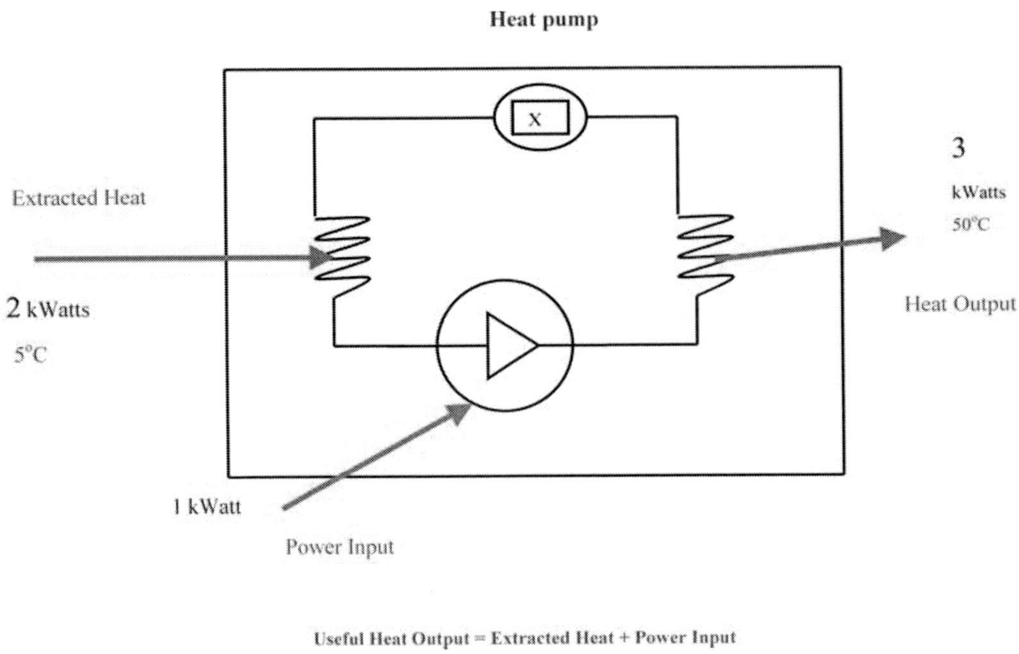

Figure 1. Schematic diagram of the GSHP

3.1. Surface Soil Heat

During the summer, solar heat is stored in the surface layer of the soil. Using this energy for heating is a practical approach for houses with a large plot (Figure 2). The amount of energy that can be extracted is greatest in soils with high water content. The heat is extracted from the soil by means of buried plastic tubing. An environment-friendly non-freezing liquid circulates in the tubing and delivers the collected heat to the heat pump. The heat pump converts the heat into high-grade heat for space heating and to produce hot water.

Figure 2. Using the soil, bedrock or groundwater as the heat source

3.2. Heat from Bedrock

Down in the bedrock there is a source of heat that stays at practically the same temperature all year round (Figure 3). Using heat from the rock is a secure, safe and environment-friendly way of heating all types of building, large and small, public and private. The capital cost is relatively high, but in return one gets a reliable, low-energy form of heating with an extremely long life. The coefficient of performance (COP) is generally good, as high as 4.8. The plant occupies little space and can even be installed on small plots. Very little reinstatement work is needed after drilling the borehole, so the effect on the nearby environment is minimal. The groundwater level is not affected, since no groundwater is used. The heat energy can be transferred to an existing, conventional, water-borne heating system and can also be used to produce hot water. The technical and economic performance of a heat pump is closely related to the characteristics of the heat source. An ideal heat source for heat pumps in buildings has a high and stable temperature during the heating season, is abundantly available, is not corrosive or polluted, has favourable thermophysical properties, and its utilisation requires low investment and operational costs. In most cases, however, the availability of the heat source is the key factor determining its use. The Table 1 presents commonly used heat sources. Ambient and exhaust air, soil and ground water are practical heat sources for small heat pump systems, while sea/lake/river water, rock (geothermal) and waste water are used for large heat pump systems.

Figure 3. Ground source heat pumps

3.3. Drilling for Energy

The heat is collected from the bedrock and the groundwater through a borehole 4.5 to 6.5 inches in diameter (Figure 4). The depth of the borehole is determined by the amount of energy needed for heating. If the energy demand is great, several boreholes can be connected together. Two lengths of tubing connected at the bottom are passed down into the borehole. Inside the collector tubing there is a frost-resistant liquid (cooling medium, and brine). The system is completely sealed, so the cooling medium never makes contact with the groundwater. To ensure that the groundwater is not contaminated by surface water running down into the borehole, a steel liner or casing which extends a short distance down into the borehole is installed. Grouting or rubber rings should be used to form a seal between casing and bedrock.

3.4. Function of the Cooling Circuit

The collector liquid (cooling medium) is pumped up from the borehole in tubing and passed to the heat pump. Another fluid circulates in the heat pump in a closed system with the most important characteristic of having a low boiling point. This fluid is called a refrigerant. When the refrigerant reaches the evaporator, which has received energy from the borehole, the refrigerant evaporates. The vapour is fed to a compressor where it is compressed. This results in a high increase in temperature. The warm refrigerant is fed to the condenser, which is positioned in the boiler water. The refrigerant gives off its energy to the boiler water, so that its temperature drops and the refrigerant changes state from gas to liquid. The refrigerant then goes via filters to an expansion valve, where the pressure and temperature are further reduced. The refrigerant has now completed its circuit and is once more fed into the evaporator where it is evaporated yet again due to the effect of the energy that the collector has carried from the energy source (Figure 5). The GSHPs come in 15 models from 4 kW up to 30 kW (even up to 300 kW when connected in parallel). At least 65% of the heating and hot water energy consumption of a house can be saved (65-75% of heating costs with a heat pump) as a result of using such a system. However, sizing of the heat pump and the ground loops is essential for the efficient operation of the system. If sized correctly, a GSHP can be designed to meet 100% of space heating requirements. The sizing of the system is very sensitive to heat loads and should therefore be installed into properties with high-energy efficiency standards, particularly new buildings. It is a good and practical idea to explore ways of minimising space heating and hot water demand by incorporating energy efficiency measures.

3.5. Free Cooling

The installation can additionally be fitted with fan convectors, for example, in order to allow connections for free cooling (Figure 6). To avoid condensation, pipes and other cold surfaces must be insulated with diffusion proof material. Where the cooling demand is high, fan convectors with drip tray and drain connection are needed.

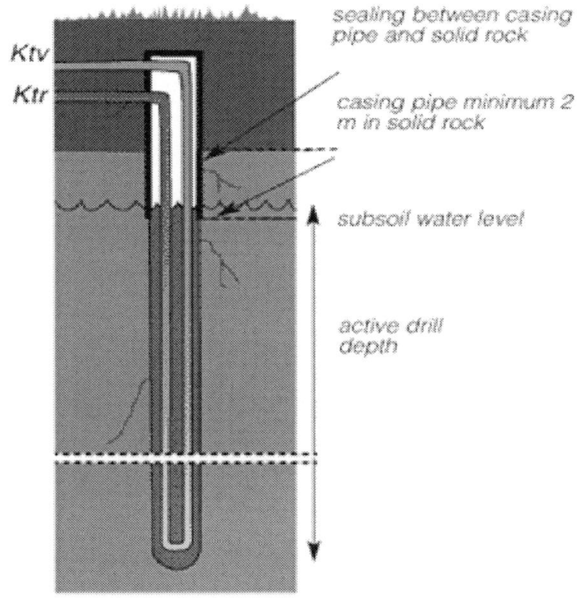

Figure 4. Borehole specifications

Ambient air is free and widely available, and it is the most common heat source for heat pumps. Air-source heat pumps, however, achieve on average 10-30% lower seasonal performance factor (SPF) than water-source heat pumps. This is mainly due to the rapid fall in capacity and performance with decreasing outdoor temperature, the relatively high temperature difference in the evaporator and the energy needed for defrosting the evaporator and to operate the fans. In mild and humid climates, frost will accumulate on the evaporator surface in the temperature range 0-6°C, leading to reduced capacity and performance of the heat pump system. Coil defrosting is achieved by reversing the heat pump cycle or by other, less energy-efficient means. Energy consumption increases and the overall coefficient of performance (COP) of the heat pump drops with increasing defrost frequency. Using demand defrosts control rather than time control can significantly improve overall efficiencies.

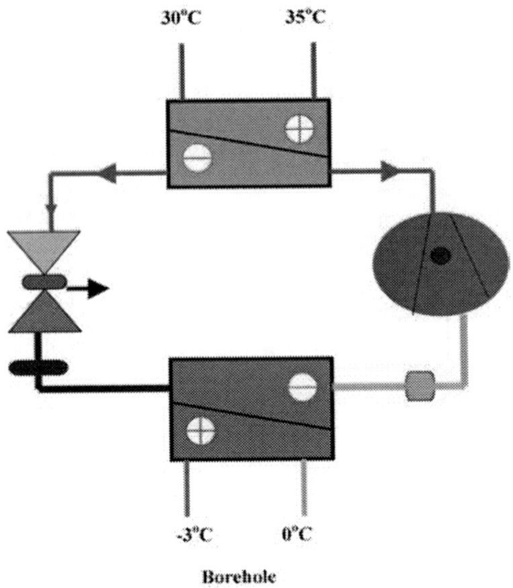

Figure 5. Function of the cooling circuit

Ground water is available with stable temperatures (4-10°C) in many regions. Open or closed systems are used to tap into this heat source. Closed systems can either be direct expansion systems, with the working fluid evaporating in underground heat exchanger pipes, or brine loop systems. Due to the extra internal temperature difference, heat pump brine systems generally have a lower performance, but are easier to maintain. A major disadvantage of ground water heat pumps is the cost of installing the heat source. Additionally, local regulations may impose severe constraints regarding interference with the water table and the possibility of soil pollution.

4. Types of the Commercial Ground Source Heat Pumps

FIGHTER 1120

Heat pumps for heating detached houses, small multi-family houses and industrial buildings using soil, bedrock or lake water as heat source.

FIGHTER 1220

Complete heat pump units for heating detached houses and terrace houses using soil, bedrock or lake water as heat source.

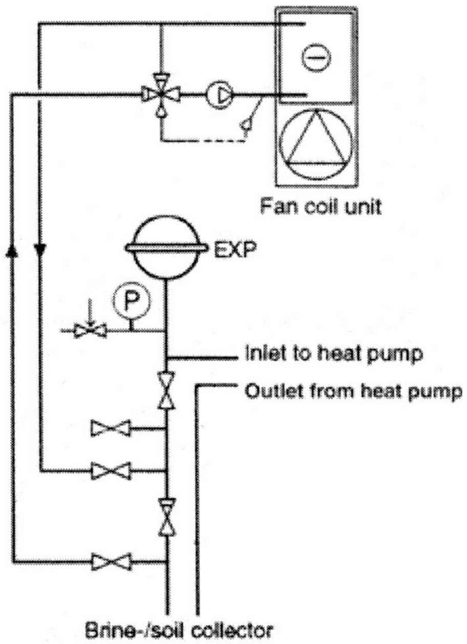

Figure 6. Diagram of cooling system

FLM 30

FLM 30 is an exhaust air module specially designed to combine recovery of mechanical exhaust air with an energy collector in the ground or in rock. FLM 30 is designed to connect to the FIGHTER 1120/1210 type of ground heat pump (Table 1). The built-in fan extracts the air from the wet areas of the house to the recovery unit. Energy is transferred to the heat pump from the ground or rock collector. The heat transfer fluid increases in temperature and enhances the heat pump heat factor. When the heat pump is not in operation, energy is stored in the ground or rock collector, which utilises the exhaust air energy to the full. The heat pump capacity is not bound by the amount of air as with an exhaust air heat pump, but the output can be optimally adapted to suit the size of the house (Table 2). FLM 30 is best connected directly to FIGHTER 1220, but can also be wall mounted. It provides a total solution for exhaust air, and ground and rock heating.

Table 1. FLM30 specifications

Height	400 mm
Width	600 mm
Depth	625 m
Net weight	60 kg
Maximum air flow	Ca 400 m^3/h
Maximum refrigerating capacity	Approximately 2 kW

Table 2. Estimation of what is possible to save with a ground source heat pump. It also provides an approximate guide for dimensioning the bore depth or hose length.

FIGHTER 1120 and FIGHTER 1220	4 [5 kW]	4 [5 kW]	5 [6 kW]	5 [6 kW]
Current total requirement, oil	2,5 m^3/year	3 m^3/year	3 m^3/year	4 m^3/year
Current total requirement, elec. *	17500 kWh/year	21000 kWh/year	21000 kWh/year	28000 kWh/year
Geothermal heat, bore depth (active) **	60 - 80 m	60 - 80 m	80 - 100 m	80 - 100 m
Ground heat, hose length **	200 - 300 m	200 - 300 m	250 - 350 m	250 - 350 m
Brine flow	0,25 l/s	0,25 l/s	0,35 l/s	0,35 l/s
Heating water flow	0,10 l/s	0,10 l/s	0,13 l/s	0,13 l/s
Saving**	11400 kWh/year	13100 kWh/year	14200 kWh/year	17700 kWh/year
	7 [9 kW]	7 [9 kW]	8.5 [10 kW]	8.5 [10 kW]
Current total requirement, oil	4 m^3/year	5 m^3/year	5 m^3/year	6 m^3/year
Current total requirement, elec. *	28000 kWh/year	35000 kWh/year	35000 kWh/year	42000 kWh/year
Geothermal heat, bore depth (active) **	110 - 130 m	110 - 130 m	130 - 160 m	130 - 160 m
Ground heat, hose length **	325 - 2x225 m	325 - 2x225 m	400 - 2x300 m	400 - 2x300 m
Brine flow	0,48 l/s	0,48 l/s	0,58 l/s	0,58 l/s
Heating water flow	0,18 l/s	0,18 l/s	0,22 l/s	0,22 l/s
Saving**	19300 kWh/year	22300 kWh/year	24500 kWh/year	28400 kWh/year
			10 [12 kW]	10 [12 kW]
Current total requirement, oil			6 m^3/year	7 m^3/year
Current total requirement, elec. *			42000 kWh/year	49000 kWh/year
Geothermal heat, bore depth (active) **			150 - 180 m	150 - 180 m
Ground heat, hose length **			2x250 - 2x325	2x250 - 2x325
Brine flow			0,65 l/s	0,65 l/s
Heating water flow			0,26 l/s	0,26 l/s
Saving**			29000 kWh/year	29000 kWh/year
Just FIGHTER 1120	11 [12 kW]	11 [12 kW]	13 [15 kW]	13 [15 kW]
Current total requirement, oil	6 m^3/year	7 m^3/year	7 m^3/year	8 m^3/year
Current total requirement, elec. *	42000 kWh/year	49000 kWh/year	49000 kWh/year	56000 kWh/year
Geothermal heat, bore depth (active) **	150 - 180 m	150 - 180 m	2x90 - 2x110 m	2x90 - 2x110 m
Ground heat, hose length **	2x250 - 2x325 m	2x250 - 2x325 m	2x300 - 2x400 m	2x300 - 2x400 m
Brine flow	0,65 l/s	0,65 l/s	0,75 l/s	0,75 l/s

Table 2 (Continued)

Heating water flow	0,31 l/s	0,31 l/s	0,31 l/s	0,31 l/s
Saving**	29000 kWh/year	32800 kWh/year	34500 kWh/year	34500 kWh/year
			15 [17 kW]	15 [17 kW]
Current total requirement, oil			8 m^3/year	9 m^3/year
Current total requirement, elec. *			56000 kWh/year	63000 kWh/year
Geothermal heat, bore depth (active) **			2x100 - 2x140 m	2x100 - 2x140 m
Ground heat, hose length **			2x350 - 3x300 m	2x350 - 3x300 m
Brine flow			0,86 l/s	0,86 l/s
Heating water flow			0,36 l/s	0,36 l/s
Saving**			39100 kWh/year	43200 kWh/year

* Total energy requirement excluding domestic electricity.

** Stated values are only approximate guide values and have been calculated based on Nordic climate conditions. A milder climate gives a greater saving.

FIGHTER 1320

A heat pump for heating larger buildings such as multi-family houses and industrial buildings using soil, bedrock or lake water as heat source.

VPA

Water heater intended primarily to be connected to heat pumps.

HPAC

An accessory that turns ground source heat pump range into a complete climate system.

4.1. Exhaust Air Heat Pumps

Ventilation, which means totally hygienic inside air, is a basic requirement for living in a healthy house. Controlled domestic ventilation with heat recovery reuses the energy from the exhaust air (Figure 7). Furthermore, the additional heat generated internally from lighting, people and domestic appliances is also utilised through heat recovery.

Figure 7. Exhaust heat pumps

In open systems the ground water is pumped up, cooled and then reinjected in a separate well or returned to surface water. Open systems should be carefully designed to avoid problems such as freezing, corrosion and fouling.

4.2. Ventilation – Health Aspects

Since the early 1980s, there has been much discussion about Sick Building Syndrome (SBS). This refers to allergic disorders, and even illness symptoms, which frequently occur in certain buildings and rooms. This can lead to chronic illness, reducing the person's ability to work and function in general. This, in turn, results, not only in the individual concerned loses his or her quality of life, but it also has a major detrimental impact on the economy and incurs huge costs. Basically, the following potential risks jeopardising people's health could be found inside buildings:

(1) Toxic pollution caused by harmful chemical substances and dust.
(2) Effects of noise, light, odours, dampness and climate.
(3) Accumulation of microbes (bacteria, viruses, mould) in terms of infection risks.
(4) Exposure to allergens.

These pollutants vary considerably according to the inside climate conditions, state of ventilation, design and use of the inside area. When energy-saving measurements were introduced in the early 1970s, considerable efforts were made to improve the insulation used in the construction industry. This led to a reduction in the air exchange rate inside buildings. From a health and allergy perspective, the ideal air exchange rate would be 0.5 – 1.0, but in actual fact, air exchange rates in appropriately insulated houses are only between 0.3 and 0.5, which means that the polluted inside air is exchanged far too infrequently. Based on the reasons given above, an increase in the incidence of complaints affecting the population health is inevitable. Significant carbon dioxide savings can be gained by displacing fossil fuels. Even compared to the most efficient gas or oil condensing boilers, a well-designed heat pump with COP of 3 to 4 will reduce CO_2 emissions by 30–35%. Further carbon savings can be made if the electricity required to power the pump comes from a renewable energy source such as photovoltaics or a renewable electricity tariff. Refrigerants such as hydrochloroflurocarbons (HCFCs) are present in the GSHP systems. These are potentially toxic, flammable and can lead to increased global warming. This is where controlled domestic ventilation can have a particular role to play. Its purpose is to control temperature and dampness, while ensuring that the quality of the inside air is totally hygienic. The relevant technical guidelines and hygiene regulations are stipulated by DIN 1946 [5].

4.3. Controlled Domestic Ventilation

Nowadays, the public generally spend around 90% of the time indoors. This undoubtedly places great demands on the climate inside. The inside climate is affected considerably by odours, harmful substances, noise and temperature.

There is a certain amount of basic ventilation in every building, even if it is only produced by air coming through windows, doors, pipe ducts and walls. This type of ventilation, in older houses in particular, provides the necessary exchange of air. Ventilation is also provided through opening windows and doors, perhaps also when one or more windows are opened at an angle. Strong wind pressure and a difference in temperature between inside and outside also increase the exchange of air. On the other hand, a weak wind or small temperature difference will reduce the required air exchange rate.

This uncontrolled ventilation also accounts for a significant part of the heating costs and causes a considerable proportion of non-renewable energy resources to be wasted. It is obvious that the ground coil can add to the cost of the system. Also many high-rise commercial applications may not have sufficient land area to accommodate a full size ground coil.

4.4. Low-Energy House

In contrast to the above, there is the low-energy house concept. The construction design is used in this type of house to prevent heat from escaping through effective thermal insulation. This also means that low-energy houses can benefit the environment as well. But, even with this construction design, there is still the problem of the required hourly air exchange rate of 0.5 – 1.0 not achieved.

To achieve the required air exchange rate either the windows would have to be opened, which would run counter to the whole low-energy house concept, or consider installing a controlled domestic ventilation system with heat recovery.

4.5. Controlled Domestic Ventilation

Controlled domestic ventilation can be used in both low-energy and older houses. In low-energy houses the controlled ventilation system guarantees the required air exchange rate, even with the doors and windows closed. When older houses are renovated better thermal insulation could be used, along with fitting new windows to enable controlled domestic ventilation to achieve the necessary air exchange rate. These types of older building are often affected by street noise. A ventilation system would therefore be beneficial in these cases too.

4.6. Controlled Domestic Ventilation with Heat Recovery

When ventilation based on opening windows and controlled domestic ventilation without heat recovery are used, the energy from the inside air is not used. The ventilation heat requirement accounts (40–50%) of the total heat requirement. In contrast to this, controlled domestic ventilation with heat recovery (Figure 8) reuses the energy from the exhaust air. Not only that, the additional heat generated internally from lighting, people and domestic appliances is also utilised through heat recovery. FIGHTER exhaust air heat pumps facilitate heat recovery and supply the energy recovered from exhaust air for the domestic hot water and even the heating. The diagram below (Figure 9) illustrates what proportion of the total heat requirement is provided by domestic ventilation with heat recovery supplied by the FIGHTER 315P and FIGHTER 600P pumps. Using the FIGHTER 600P, it is possible to supply a far greater proportion of the total heat requirement with the heat pump. In this case, energy recovery ensures a healthy and comfortable form of heating, while the system also produces considerable savings in terms of heat energy, along with CO_2 emissions.

Geothermal energy can be used to produce steam for electricity generation, hot water for use in horticulture, fish farming, etc., or heat for individual buildings and district heating schemes. Opportunities vary with the temperature of the water extracted. For instance, water from aquifers at temperatures in the 50-150°C range has applications in district heating, fish farming, horticulture and recreational facilities such as spaces.

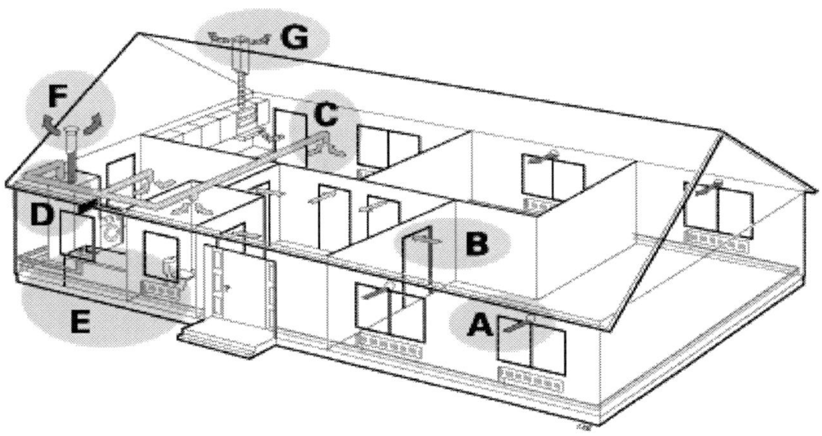

Figure 8. Domestic ventilation

Functions

A Fresh outside air is supplied to the house through cleanable outside vent holes.
B The air overflow occurs under the door or through the overflow vent holes.
C The warm inside air (exhaust air) is drawn into the ventilation system.
D Warm exhaust air is supplied to the heat pump for heat recovery.
E The heat pump provides the house with domestic hot water and/or hot water for the radiators.
F When the exhaust air has passed through the heat pump the discharge air is released into the open air.
Before this, the heat pump has extracted so much energy from the exhaust air that the temperature of the discharge air is only about 0°C (depending on the system).
G Extractor hood.

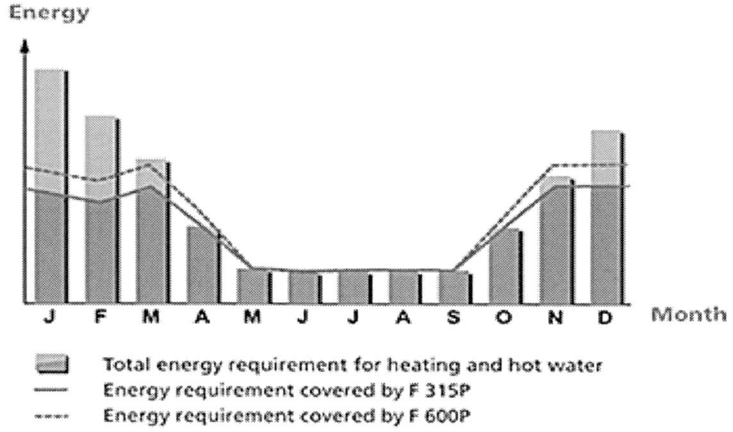

Figure 9. Energy requirements for heating and hot water

5. Sizing Considerations

Sizing of the heat pump and the ground loop is crucial to the operation of the system and will depend on the heating requirements. At the outset it is a good idea to look at energy efficiency measures such as wall and floor insulation to reduce the heat demand. When sizing a system, it is important to consult a professional installer for expert advice. A heat pump that can be designed to meet 100% of space heating requirements is usually only used to pre-heat domestic hot water, so top-up heating (e.g., an immersion heater) would also be required.

Sizing of the heat pump and the ground loops is essential for the efficient operation of the system. If sized correctly a GSHP can be designed to meet 100% of space heating requirements. However, the sizing of a system is very sensitive to heat loads and should therefore be installed into properties with high-energy efficiency standards, particularly new buildings. Also, it is always a good idea to explore all possible ways of minimising space heating and hot water demand by incorporating energy efficiency measures.

Unlike the outside air, the temperature of the ground remains fairly constant. As a result, the potential output of an Earth Energy System (EES) varies little throughout the winter. Since the EES's output is relatively constant, it can provide almost all the space heating requirement - with enough capacity left to provide hot water heating as an "extra". As with air-source heat pump systems, it is not generally a good idea to size an EES to provide all of the heat required by a house. For maximum cost-effectiveness, an EES should be sized to meet 60 to 70 percent of the total maximum "demand load" (the total space heating and water heating requirement). The occasional peak-heating load during severe weather conditions can be met by a supplementary heating system. A system sized in this way will in fact supply about 95 percent of the total energy used for space heating and hot water heating.

EESs with variable speed or capacity are available in two speed compressor configurations. This system can meet all cooling loads and most heating loads on low speed, with high speed required only during high heating loads. A variety of sizes of EESs, particularly those suits the Canadian climate, are available in he market. Units range in size from 0.7 kW to 35 kW (2400 to 120000 Btu/h), and include domestic hot water (DHW) options.

Three options are available for the ground loop: borehole, straight horizontal and spiral horizontal (or 'slinky'). Each has different characteristics allowing the choice of the most suitable option for the property. Horizontal trenches can cost less than boreholes, but require greater land area. For slinky coil, a trench of about 10 m long will provide for about 1 kW of heating load. There are three important elements to a GSHP system: ground loop, heat pump and heat distribution system. These are described below.

5.1. Ground Loops

Lengths of plastic pipe are buried in the ground, either in a borehole or a horizontal trench. The pipe is a closed loop, which is filled with a water/antifreeze mixture. This mixture circulates in the pipe, absorbing heat from the ground.

Horizontal trenches are drilled to a depth of 1 to 2 metres and can cost less than boreholes, but require a greater area of land. Placing coiled piping in horizontal trenches will enhance the performance compared with straight piping.

A borehole is drilled to a depth of between 15 to 100 metres and will benefit from higher ground temperatures than the horizontal trench, although installation costs will be greater.

5.2. Heat Pumps

The heat pump works by promoting the evaporation and condensation of a refrigerant to move heat from one place to another. A heat exchanger transfers heat from the water/antifreeze mixture in the ground loop to heat and evaporate refrigerants, changing them to a gaseous state.

A compressor is then used to increase the pressure and raise the temperature at which the refrigerant condenses. This temperature is increased to approximately 40°C. A condenser gives up heat to a hot water tank, which then feeds the distribution system.

5.3. Heat Distribution Systems

Because the GSHPs raise the temperature to approximately 40°C they are most suitable for underfloor heating systems, which require temperatures of 30 to 35°C, as opposed to conventional boiler systems, which require higher temperatures of 60 to 80°C. The GSHPs can also be combined with radiator space heating systems and with domestic hot water systems. However, top-up heating would be required in both cases in order to achieve temperatures high enough for these systems. Some systems can also be used in reverse mode for cooling in the summer.

5.4. Design Considerations

As explained earlier, unlike air-source heat pumps, EESs require that a well or loop system be designed to collect and dissipate heat underground.

5.5. Earth Loops

A constant ground temperature of about 53 degrees Fahrenheit all year long a few metres below the surface, such as in Lane County, USA, makes an ideal source for heating and cooling buildings with a GSHP. Because of this relatively constant temperature of the source, the soil, all year-round, a GSHP has several advantages over an air-source heat pump. It is highly efficient with a COP of about 3.5 to 4. It uses much less supplemental resistance heat. It does not require the energy sapping defrost cycle. Currently, used design depths result in minimum brine temperatures in the range –3 to -5°C. This temperature can be raised by means of a deeper borehole or recharging from an alternative heat source (e.g., exhaust-air or a solar system).

5.6. Types of Earth Loops

Various types of earth loops are discussed below.

5.6.1. Open loop

The simplest GSHP installation uses ground water as a direct source of heat. Water is pumped from a well, circulates through the heat pump and is discharged to the surface. Open loops require good quality well water at a reasonable pumping depth. The well must have enough capacity to provide a continuous flow of 1.5 gallons per minute per ton of heat pump capacity during peak operation. An approved discharge system must be available. Ditches, field tiles, ponds, streams and dry wells are the most common methods, although local regulations may limit the use of this approach. Open loop systems are usually less expensive than other configurations, but restrictions on the way water is discharged and concern about the large quantity of water used make other approaches attractive to many people. As noted, an open system (Figure 10) uses ground water from a conventional well as a heat source. The ground water is pumped into the heat pump unit, where heat is extracted. Then, the "used" water is released in a stream, pond, ditch, drainage tile, river, or lake. This process is often referred to as the "open discharge" method.

Another way to release the used water is through a rejection well, which is a second well that returns the water to the ground. A rejection well must have enough capacity to dispose of all the water passed through the heat pump, and should be installed by a qualified well driller. Regardless of the approach used, the system should be designed to prevent any environmental damage. The heat pump simply removes or adds heat to the water; no pollutants are added. The only change in the water returned to the environment is a slight increase or decrease in temperature.

The size of the heat pump unit and the manufacturer's specifications will determine the amount of water that is needed for an open system. The water requirement for a specific model of heat pump is usually expressed in litres per second (L/s) and is listed in the specifications for that unit. The average heat pump unit of 10 kW (34000 Btu/h) capacity tends to use 0.45 to 0.75 L/s while operating.

The well and pump combination should be large enough to supply the water needed by the heat pump in addition to the domestic water requirements. One may need to enlarge the normal domestic pressure tank or modify the plumbing to supply adequate water to the heat pump.

The variable speed air unit and the ground-source heat pump are well suited to meet the dehumidification need. The constant speed unit has poor latent cooling capacity and discomfort is a possibility during mild cooling days. Winter air delivery temperatures with the ground machines will be 98 to 102°F with the heat recovery unit operational and 100 to 106°F if it is off. The air machines will deliver air at a lower temperature. The ground-source machine had lower demand (summer and winter) and lower heating energy use than either of the air heat pumps.

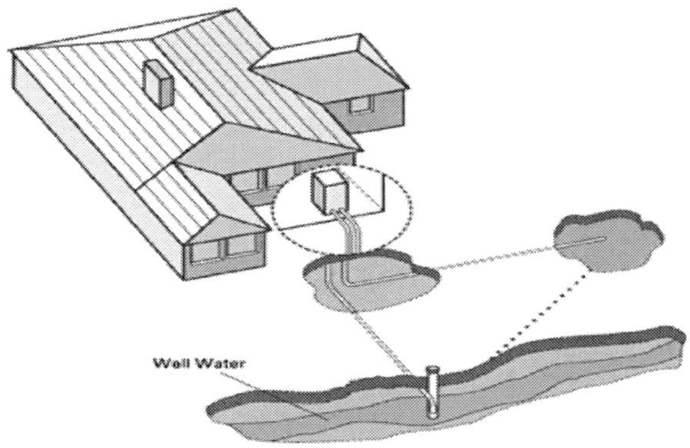

Figure 10. Open system using ground water from a well as a heat source

Poor water quality can cause serious problems in open systems. As a rule of thumb, one should not use water from a spring, pond, river, or lake as a source for the heat pump system unless it has been proven to be free of excessive particles and organic matter, and warm enough throughout the year (typically over 5°C) to avoid freeze-up of the heat exchanger. Particles and other matter can clog the heat pump system and make it inoperable in a short period of time. Water should also be tested for acidity, hardness, and iron content before installing the heat pump to ensure meeting the acceptable level of water quality as required by the heat pump and heat exchanger materials specifications. Additionally, the installation of an open system is often subject to local zoning laws or licensing requirements that need to be satisfied as well.

5.6.2. Closed loop

Closed loop systems use a circuit of sealed pipe to exchange heat with the ground. While an open system drains water from a well, a closed-loop system recalculates its heat transfer solution in pressurised pipe. The pipe itself is high-density polyethylene plastic with an expected service life of 50 to 75 years. The pipe is connected to the indoor heat pump to form a sealed underground loop through which an antifreeze solution or refrigerant is circulated. Sections of pipe are joined with heat fusion for a reliable seal. However, copper tubing is used in the case of DX systems. A small pump circulates a mixture of water and non-toxic antifreeze (usually propylene glycol) through the pipe. This "earth loop" exchanges heat with the surrounding soil.

The length of loop depends on the size of the heat pump equipment, the moisture content of the soil and the climate. For example, the typical earth loop is 110 linear feet per ton [6]. When space heat is required, the heat pump removes heat from the soil. In cooling mode, the heat pump rejects heat to the soil. The large "thermal mass" of the soil can easily absorb these temperature swings.

The pipe is placed in one of two types of arrangements: vertical or horizontal. A vertical closed-loop arrangement (Figure 11) is an appropriate choice for most suburban homes, where lot space is restricted. Piping is inserted into bored holes that are 150 mm (6 in) in diameter, to a depth of 18 to 60 m (60 to 200 ft), depending on soil conditions and the size of the system. Usually, about 80 to 110 m (270 to 350 ft) of piping is needed for every ton (3.5

kW or 12000 Btu/h) of heat pump capacity. U-shaped loops of pipe are inserted in the holes. DX systems can have smaller diameter holes, which can lower drilling costs.

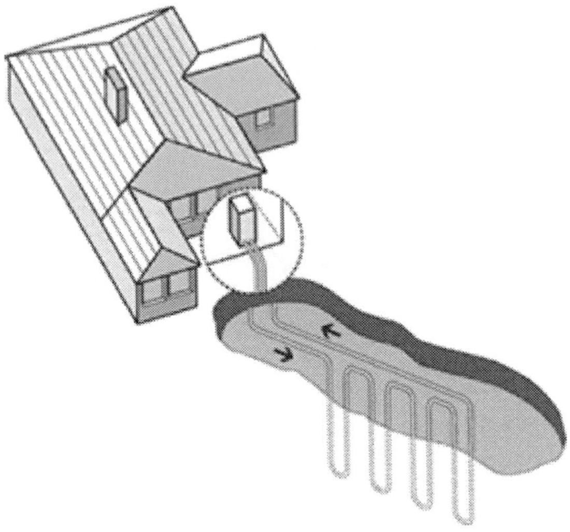

Figure 11. Closed-loop, single U-bend vertical configuration

5.6.3. Pond loops

Pond loops reduce excavating costs by placing much of the loop in a pond, lake or stream. In most cases, 1/4 to 1/2 acre of water surface and a minimum depth of six to eight feet are required.

5.6.4. Vertical loops

Vertical loops require fewer pipes than horizontal loops. Well-drilling equipment bores small diameter holes 50 to 200 feet deep. Two pipes joined with a U-bend are inserted into each hole. The system can use one deep hole or several shallow ones. Verticals can be installed almost anywhere. Because they require only 100 to 200 square feet per ton, they are ideal for small lots. However, the cost of drilling is often greater than trenching.

5.6.5 Horizontal loops

The earth loop exchanges heat with the soil. Horizontal loops are most common (Figure 12). A four to six feet deep trench is dug using a backhoe or chain trencher. The pipe is laid out, sealed and pressure is tested. Then the trench is filled. The land area required ranges from 750 to 1,500 square feet per ton.

This horizontal arrangement (Figure 13) is more common in rural areas, where properties are larger. The pipe is placed in trenches normally 1.0 to 1.8 m (3 to 6 ft) deep, depending on the number of pipes in a trench. Generally, 120 to 180 m (400 to 600 ft) of pipe is required per ton of heat pump capacity. For example, a well-insulated, 185 m^2 (2000 ft^2) home would probably need a three-ton system with 360 to 540 m (1200 to 1800 ft) of pipe.

Figure 12. Earth loop exchanges heat with the soil

The most common horizontal heat exchanger is two pipes placed side-by-side in the same trench. Another heat exchanger sometimes used where area is limited is a "spiral", which describes its shape. Other horizontal loop designs use four or six pipes in each trench, if land area is limited. Regardless of the arrangement, all piping for antifreeze solution systems must be at least series 100 polyethylene or polybutylene with thermally fused joints (as opposed to barbed fittings, clamps, or glued joints) in order to ensure leak-free connections for the life of the piping. Properly installed, these pipes will last between 25 to 75 years. They are unaffected by chemicals found in soil and have good heat-conducting properties. The antifreeze solution must satisfy local environmental requirements. DX systems use refrigeration-grade copper tubing. Neither the vertical nor the horizontal loops have an adverse impact on the landscape as long as the vertical boreholes and trenches are properly backfilled and tamped (packed down firmly).

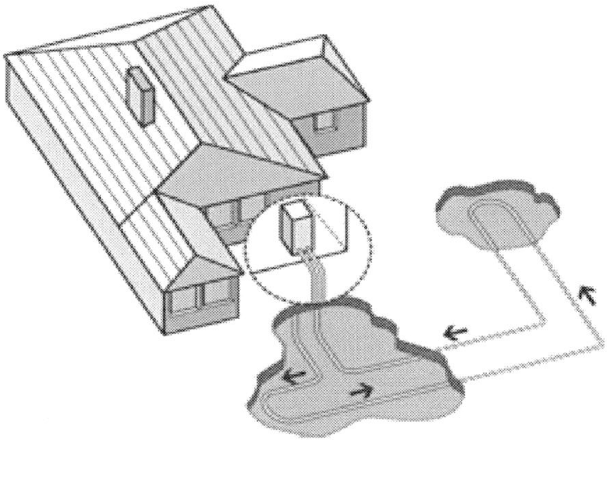

Figure 13. Closed-loop, single layer horizontal configuration

Horizontal loop installations use trenches between 150 to 600 mm (6 to 24 in) wide. This leaves bare areas that can be restored with grass seed or sod. Vertical loops require little space and result in minimal lawn damage. It is important that a qualified contractor install the horizontal and vertical loops. Plastic piping must be thermally fused, and there must be good earth-to-pipe contact to ensure good heat transfer, such as that achieved by Termite-grouting of boreholes. The latter is particularly important for vertical heat-exchanger systems. Improper installations may result in less than optimum heat pump performance.

5.6.6. Vertical closed loop

In a vertical closed-loop ground heat exchanger, a water/antifreeze mixture is circulated through sealed pipe loops buried in vertical boreholes. The boreholes are typically 150 to 200 feet deep. As with a horizontal closed-loop system, heat is transferred by the heat pump system, from the ground during the winter and to the ground during the summer. A vertical heat exchanger can be installed on smaller lots in preference to a horizontal system (Figure 14).

5.6.7. Horizontal closed loop

In a horizontal closed-loop ground heat exchanger, a water/antifreeze mixture is circulated through sealed pipe loops buried horizontally, about six feet underground. During cold weather, the pipe loops absorb heat from the earth and deliver it to the heat pump located in the house. The heat pump transfers heat from the loop to warm the air that is circulated throughout the house by ductwork. The special, non-toxic antifreeze ensures that the system will not freeze during severe winter weather. In the summer, the process is reversed for air conditioning, and the heat pump system transfers heat from the house to the ground (Figure 15).

The potential for improvement (in terms of performance and lower costs) must be evaluated thoroughly before the industry's limited research resources are unwisely expended. Alternatives should have the following characteristics:

1. Very durable during and after installation.
2. Able to be installed with low cost, low maintenance, and compact equipment.
3. Require moderate volumes of antifreeze solutions in cold climates.
4. Must not require heat pumps with larger refrigerant charges, complicated controls or complex refrigerant circuits.
5. Be evaluated based on long periods (one year) with higher than average loads.
6. Constructed of materials that can be fabricated with conventional materials that will not decay in soils.
7. Able to be installed and serviced by technicians with moderate skills.

It is obvious that the ground coil can add to the cost of the system. Also many high-rise commercial applications may not have sufficient land area to accommodate a full size ground coil. A hybrid ground-coupled water-loop heat pump (GCWLHP) would be a viable option to reduce the size of the ground coil. The coil would be sized to meet the heating requirement of the building. This is typically one-half the size required for meeting the cooling load. There are several reasons for the smaller size:

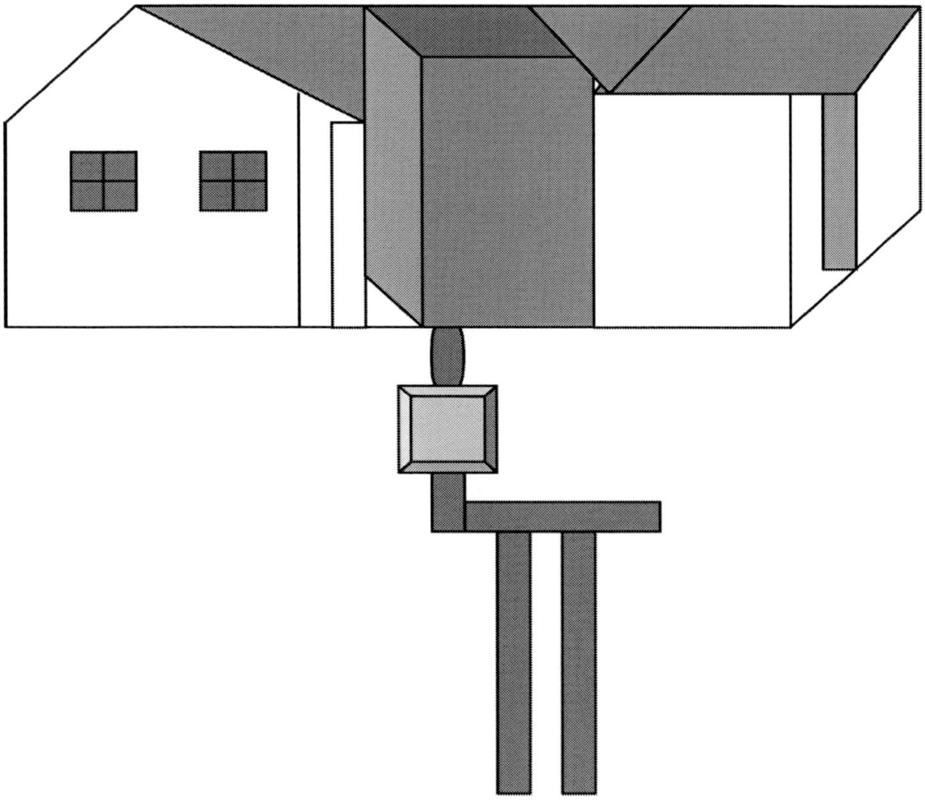

Figure 14. Vertical closed-loop

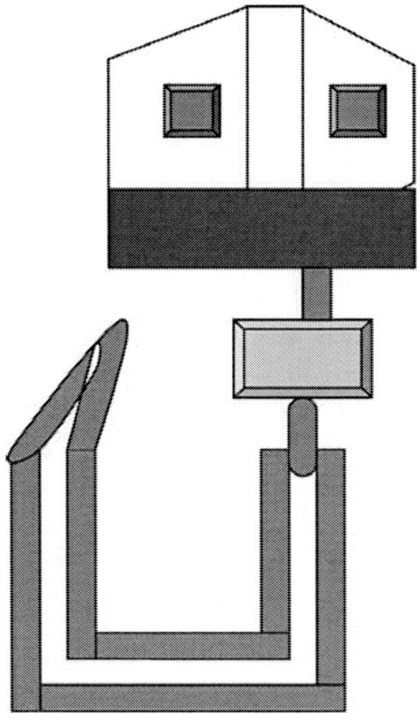

Figure 15. Horizontal closed-loop

6. Indoor Equipment

The GSHPs are similar to ordinary air conditioners and air-source heat pumps, but they dump building heat into or extract heat from the ground rather than the outside air. The tempering effect of the earth provides for either heat extraction or heat rejection, allowing the GSHPs to operate more efficiently than conventional equipment in particularly cold or hot climates.

All GSHP systems consist of a "ground loop" for exchanging heat with soil, groundwater, surface water, or wastewater streams. In most systems, the heat exchanger consists mainly of long pipes either placed in vertical bores or, if sufficient land is available, buried in horizontal trenches. One or more heat pumps are then connected to this ground loop to provide indoor space heating and cooling. Some systems provide water heating as well. The ground loop is most commonly a closed loop system, but some open loop systems are used whereby water is pumped through the GSHP and then re-injected into the ground or into surface water.

The indoor equipment for a ground-source heat pump takes up about as much space as a typical furnace (Figure 16). Most of the equipment is located inside the home. The main unit, which looks like an electric furnace, holds the compressor, blowers, heat exchange coil and control. The only additional equipment needed is the pump that circulates fluid through the earth loop. The domestic hot water tank usually sits nearby, because the GSHP also generates an ample supply of hot water. The GSHPs offer an efficient and reliable alternative for home comfort system.

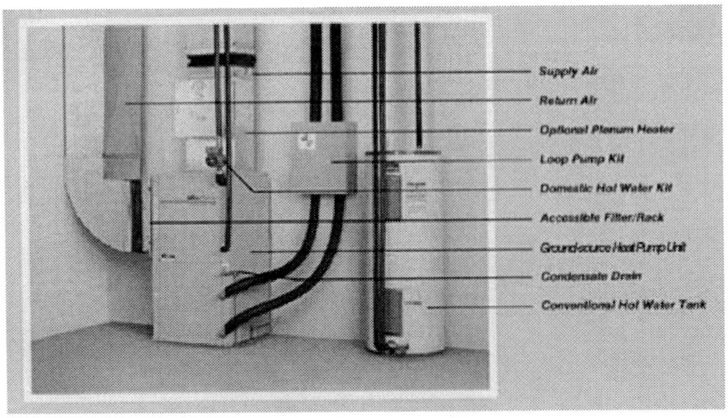

Figure 16. Indoor equipment

7. Applications

The GSHPs can be applied to virtually any size residential buildings as well as specialised applications such as swimming pools. The vertical ground loops average 400 feet in length per tone of capacity; so limited site area is usually not a problem. Horizontal loops may be 600 to 1000 feet per ton, depending on soil type. The GSHPs tend to be the most cost effective in the following situations:

- New construction.
- Climates characterised by high daily temperature swings, or where winters are quite cold or summers quite hot.
- Areas where electricity costs are over $.07/kWh.
- Areas where natural gas is unavailable or where cost is high compared with electricity.

When considering an open-loop system, check with local environmental authorities to ensure compliance with regulations. There are one-to-two-million GSHPs are operating in homes, schools, and commercial buildings in the United States. They are adaptable to virtually any kind of building; the federal government has installed many thousands of the GSHPs, mostly in residential applications [6].

7.1. Basic Description of the Component of a GSHP

The ground heat source: Is usually a closed loop of plastic pipe containing a glycol antifreeze solution. The pipe is buried in the ground in vertical boreholes or horizontal trenches. The trenches take either straight pipe or coiled (slinky) pipe, buried about 1.5 to 2 m below the surface. A large area is needed for this application.

Heat pump: A device that transfers heat from a lower temperature reservoir to a higher temperature reservoir by doing work through a compressor.

A heat pump packaged unit: Water-Water type (Approximately the size of a small fridge) containing two cold water (glycol) and two heated water connections.

Open loop: A heating or cooling system, such as a solar water heater or ground source heat pump, in which the working fluid is heated or cooled and used directly, not returning to the heating device.

The heat distribution system: This is either underfloor heating pipes or conventional radiators of large area connected via normal water pipes.

Electrical input and controls: The system will be requiring an electrical input energy, three-phase being preferred, but single phase is perfectly adequate for smaller systems. A specialised controller will be incorporated to provide temperature and timing functions of the system.

7.2. Performance and Costs

The GSHPs energy cost savings vary with the electric rates, climate loads, soil conditions, and other factors. In residential building applications, typical annual energy savings are in the range of 30 to 60 percent compared to conventional HVAC equipment. Unlike air source units, GSHP systems do not need defrost cycles nor expensive backup electric resistance heat at low outdoor air temperatures. The stable temperature of a ground source is a tremendous benefit to the longevity and efficiency of the compressor.

A residential GSHP system is more expensive to install than a conventional heating system. It is most cost-effective when operated year-round for both heating and cooling. In such cases, the incremental payback period can be as short as 3–5 years. A GSHP for a new residence will cost around 9-12% of the home construction costs. A typical forced air furnace with flex ducting system will cost 5-6% of the home construction costs. Stated in an alternative form, the complete cost of a residential GSHP system is $3,500-$5,500 per ton [6]. Horizontal loop installations will generally cost less than vertical bores. For a heating dominated residence, figure around 550 square feet/ton to size the unit. A cooling dominated residence would be estimated around 450 square feet/ton. The table 3 below compares three types of systems.

Table 3. Comparisons of central heating systems

Central system type	Concern for safety	Installation cost	Operating cost	Life-cycle cost
Combustion-based	Some concern	Moderate	Higher	Moderate
Heat pump	Less concern	Moderate	Low	Moderate
GSHP	Less concern	High	Lower	Moderate to low

8. CONCLUSIONS

Conserving natural resources benefits everyone now and into the future. For homebuilders, green building means the resource-efficient design, construction, and operation of homes. It represents an approach to both building and marketing homes that highlights environmental quality.

The GSHPs provide an effective and clean way of heating buildings worldwide. They make use of renewable energy stored in the ground, providing one of the most energy-efficient ways of heating buildings. They are suitable for a wide variety of building types and are particularly appropriate for low environmental impact projects. They do not require hot rocks (geothermal energy) and can be installed in most of the world, using a borehole or shallow trenches or, less commonly, by extracting heat from a pond or lake. Heat collecting pipes in a closed loop, containing water (with a little antifreeze) are used to extract this stored energy, which can then be used to provide space heating and domestic hot water. In some applications, the pump can be reversed in summer to provide an element of cooling. The only energy used by the GSHP systems is electricity to power the pumps. Typically, a GSHP will deliver three or four times as much thermal energy (heat) as is used in electrical energy to drive the system. For a particularly environmental solution, green electricity can be purchased. The GSHP systems have been widely used in other parts of the world, including North America and Europe, for many years. Typically they cost more to install than conventional systems; however, they have very low maintenance costs and can be expected to provide reliable and environmentally friendly heating for in excess of 20 years. Ground source heat pumps work best with heating systems, which are optimised to run at a lower water temperature than is commonly used in the UK boiler and radiator systems. As such, they make an ideal partner for underfloor heating systems.

REFERENCES

[1] Mortal, A., (2002). *Study of solar powered heat pump for small spaces.* Portugal.

[2] Shao, S., et al. (2002). *Thermodynamic analysis on heat pumps with economiser for cold regions*. China.

[3] Yamakamika, O. (Japan. 1998). Development of burner for room air conditioner. *Technical Review Mitsubishi Heavy Industries*, Vol.*35*.

[4] Horiuchi, N. (1997). Development of packaged air conditioner for cold regions. *Refrigeration,* Vol.*72*, No.*837*.

[5] Roriz, L. (2001). Determining the potential energy and environmental effects reduction of air conditioning systems. *Commission of the European Communities DG TREN.*

[6] Witte, H., et al. (2002). *Comparison of design and operation of a commercial UK ground source heat pump project.* Groenholland BV.

In: Energy Costs, International Developments...
Ed: Leszek Kowalczyk and Jakub Piotrowski
ISBN: 978-1-60741-700-2

Chapter 6

ENERGY EFFICIENT USE IN THE PLANNING AND MANAGEMENT OF WATER DISTRIBUTION SYSTEMS

Inmaculada Pulido-Calvo* and Juan Carlos Gutiérrez-Estrada
Dpto. Ciencias Agroforestales, EPS, Campus Universitario de La Rábida, Universidad de Huelva, 21819 Palos de la Frontera (Huelva), Spain

ABSTRACT

The energy cost is one of the most important cost components in the water supply systems. Since large amounts of electricity are required to pump, transport and apply water, the profitability of some business, as irrigation districts or fishfarms, which use the water as production resource, is heavily dependent upon energy costs. Methodologies that can maximize energy cost savings while satisfying system performance criteria should be sought for the design and management of the water distribution systems. Some of these methods can be: (a) to improve the selection and/or operation of pumping stations; and (b) to include a regulating reservoir between the water supply source and the delivery system.

The alternative (a), related with the optimal design of pumping stations, refers to the selection of pump type, capacity and number of pump groups that results in minimum design and operating costs for a given water demand. With the alternative (b), if time-of-day energy tariff is available, the reservoir can be used for storing water that is pumped during off-peak hours to be used during peak hours. Also, the inclusion of reservoirs allows an equalization of pumping schemes that is not possible by pumping directly into the water delivery system. This allows pumping stations to operate near optimum energy efficiency.

In this chapter, one model that includes both alternatives was presented. To implement this optimization process, two algorithms were developed: (a) Algorithm for selection of least cost or optimum pump combinations in water supply systems and to evaluate the system's energy costs; and (b) Algorithm to determine the reservoir storage capacity that permits water to be pumped when energy tariffs are lowest and establish an annual pumping schedule in accordance with time-of-use energy tariffs. This model was applied to the water distribution systems of one irrigation district and one fishfarm

* Correspondence author: ipulido@uhu.es (I. Pulido-Calvo), juanc@uhu.es (J.C. Gutiérrez-Estrada)

located in southern Spain. The analysis indicated that the optimal selection of pump groups and/or the addition of a regulating reservoir was/were energy cost effective.

1. INTRODUCTION

The rapid increase in energy prices that has occurred during the last decades has created the need for increased emphasis on efficient energy use. In many water distribution systems, due to large amounts of energy are required to pump, transport and apply water, improved management of pumps leading to a reduction in energy usage and operational cost must therefore be regarded as a priority when more efficient network operation is sought. In solving this problem, account should be taken of the efficiencies of the pumps, the structure of the electricity tariff, the consumer-demand pattern, and the possibility of a regulation reservoir in the system. The interaction between the pump controls, the resulting pump power consumptions, and the energy head and flow regime in the network will have to be considered through the nonlinear network hydraulics and pump characteristics.

It can be seen that the problem of optimal pump scheduling is one of high complexity for which a formal approach is required. The formulation of the optimization problem must be such that sufficient representation of the network hydraulic characteristics and pumping costs is included, without the resulting solution being too complex for its computer implementation. The need for an efficient solution is emphasized by the fact that the rapidly varying nature of consumer demands requires pump schedules to be obtained in real time for the full benefits of optimal control to be achieved (Jowitt and Germanopoulos, 1992).

As a result, simplifying assumptions have been made to reduce the dimensionality and complexity of the optimal pump scheduling problem in urban water supply systems (Quimpo and Shamsi, 1991). The methods ranged from heuristic methods (Walski, 1984; Tarquin and Dowdy, 1989), to mathematical modeling methods. These methods have been studied using several techniques such as linear programming (Jowitt and Germanopoulos, 1992; Crawley and Dandy, 1993); dynamic programming (Sabet and Helweg, 1985; Ormsbee *et al.*, 1989; Zessler and Shamir, 1989; Biscos *et al.*, 2003) and nonlinear programming (Brion and Mays, 1991; Cembrano *et al.*, 2000). Hierarchical-descomposition methods were also suggested (Joalland and Cohen, 1980; Nitivattananon *et al.*, 1996). Another alternative has been the application of expert systems that combine heuristic procedures with algorithmic calculations (Shepherd and Ortolano, 1996; León *et al.*, 2000).

In the case of irrigation water supply systems, a number of studies have suggested methods to reduce energy costs for pumping. Buchleiter and Heermann (1986, 1990), Moradi-Jalal *et al.* (2003) and Planells *et al.* (2005) have developed methods to optimize the type and number of pumps as well as scheduling the operation of irrigation pumps, considering both the initial investment and the cost of consumed energy. Stetson *et al.* (1975) showed that significant peak electrical demand reduction could be achieved by shifting irrigation to off-peak hours where pumping plants have sufficient capacity to replace the water consumed by the crop. Some power suppliers have implemented load shedding programs to reduce the operating load on the transmission and delivery system during periods of peak demand. Typically, these programs result in irrigation pump shutoff on a predetermined rotational schedule and get savings in the form of discounts on the power bill to the irrigators. This way, authors like Buchleiter *et al.* (1984), Duke *et al.* (1984) and

Heermann *et al.* (1984) made an automatic control system which sets priorities for load interruption in irrigation pump based on soil and crop water status.

The main goal of the most of the aforementioned investigations in urban and irrigation water supply systems is to define the optimal scheduling of the pumping station over a 24-h period. For every hour, the solution must identify the pump, or pump combination, which should be working in order to satisfy the water demand at minimum cost. Nearly all of these works are limited to the operation phase of a given or defined water supply system. In the case of the inland intensive fishfarms, in spite of the high energy costs required for pumping, it is surprising to find few studies in the literature based on the optimum design and management of the water supply systems (Kerr, 1981; Pulido-Calvo *et al.*, 2006a, 2008).

Major difficulties in the optimization of water supply system operation include the following: (a) Size and configuration differences among water distribution systems. Large computational time and memory requirements occur in a large system and long planning period (most of the existing models for water supply system operation only assume a 24-h water demand pattern-short planning period-), and; (b) The discrete pump discharges. When pump speeds are considered fixed, the solutions for pump discharges are a discrete set of feasible operating points (Nitivattananon *et al.*, 1996).

In this chapter, hourly demand histograms during the annual operation period were considered (long planning period). The full demand during the annual operation period must be satisfied through the designed pumping station. The annual water demand curve allows to select pump combinations which result in the lowest annual total cost (annual depreciation cost and annual operation cost), and that the energy cost evaluation of the delivery system is determined in a more approximate form in comparison with the 'traditional' methods that consider only the point of maximum necessity operation (Stetson *et al.*, 1975; Lansey and Mays, 1989; Jowitt and Germanopoulos, 1992; Breytenbach *et al.*, 1996).

The construction of a regulation reservoir between the source of water supply and the distribution network is frequent in the water distribution systems. This system is motivated by the need to narrow the temporal gaps between supply and demand and thus to enhance the degree of freedom in water consumers (Mehta and Goto, 1992; Nel and Haarhoff, 1996; Hirose, 1997; Pulido-Calvo *et al.*, 2006b). Therefore, the construction of a regulation reservoir or tank may reduce the operational cost of water supply systems because is used for storing water that is pumped from wells or other sources of supply during off-peak periods when energy costs are less for use during periods of peak electrical demand with high energy costs (Sabet and Helweg, 1989).

This chapter presents the development and application of an optimization model that considers the aforementioned difficulties. The objective is to choose pump combinations and a storage capacity that allow to elevate the water in the hours with the most advantageous energy cost and to establish an operation of pumping adapted with the time-of-use energy tariff while satisfying water demands and system hydraulic requirements. The optimal storage capacity is that which results in the minimum total cost, which includes the annualized value of the capital cost of the system (pumps and reservoir) and the operating costs (energy cost). The determination of an optimum pumps and reservoir design and long-term optimum operation rule for pumping to the reservoir is of interest in the model developed. Integration of the design phase with the operation schedule in an optimization method is the main purpose of this paper.

Once the optimization process and the method solution were identified, a program for a personal computer was written to find the optimal design and operation of the pumping stations and/or regulating reservoir. This program was named DYGOSIA v.1.0 and was written in Microsoft Visual Basic® programming language. The mathematical model development in this chapter forms the basis of DYGOSIA v.1.0 software. The model was verified by applying the developed software to two existing water distribution systems in an irrigation district and in an inland intensive fishfarm, located in the southern Spain.

2. Formulation of Objective Function

The model described in this chapter minimizes an objective function equal to the sum of the equivalent annual value of the capital cost C_I (in €) for T years when present capital cost is invested at r percent interest (fixed cost = depreciation cost of pumps and reservoir) and the operating cost (variable cost = energy cost):

$$\min i \min ize\ C_{total} = \left\{ \sum_{t=1}^{NE} \sum_{p=1}^{nb} \left[\frac{\gamma\, Q_p(t)\, H_p(t)}{\eta_p(t)} C_{Ep}(t)\Delta t \right] + \frac{r\,(1+r)^T}{(1+r)^T - 1} C_I \right\} \tag{1}$$

where C_{total} is the total annual cost; NE is the number of time steps of the optimization procedure; nb is the number of pumps; γ is the specific gravity of the water; $Q_p(t)$ is the p pump discharge during the time step t; $H_p(t)$ is the discharge energy head of the p pump at the time step t; $\eta_p(t)$ is the p pump efficiency at the time step t; $C_{Ep}(t)$ is the energy cost of the p pump during the time step t, in €/kWh; and Δt is the length of the time step t.

This objective function is subject to:

- Limitations on pump discharge $Q_p(t)$ and discharge energy head $H_p(t)$ that are functions of the hydraulic characteristics of the pumps and their operation schemes.
- Max-min reservoir volumes:

$$V_{min} \le V(t) \le V_{max} \qquad \forall t \tag{2}$$

  where V_{min} and V_{max} are the minimum and maximum useful regulating capacities, respectively, and V(t) is the useful stored volume in reservoir in the time step t.
- Mass balance, that is to say, the difference of the useful stored volume in reservoir between the time steps t and t-1 must be equal to the difference between the pump discharge $Q_p(t)$ and the water demand Demand(t) of the delivery system at time step t:

$$V(t) - V(t-1) = \left[Q(t) - \text{Demand}(t) \right] \Delta t \qquad \forall t \tag{3}$$

The optimization period is divided into a discrete control hourly intervals because the demands display a pronounced daily cycle, and energy tariffs are based on time of day.

3. Model Analysis and Solution

There are two difficulties in formulating this model. First, the cost function tends to be non-linear in terms of pump discharge and discharge energy head. Second, a computational time problem occurs when a large number of time steps for a planning period is considered in the model. As a result, an iterative solution of the optimization problem is generally required. The preparation for optimization includes setting up a data structure to manipulate and process the following: Water supply system configuration; Water demand data; Structure of electricity tariff; Data files of pump groups and reservoirs. The steps in the development of the optimization model can be explained as follows:

(a) Decomposition

There are two types of decomposition, including:

- Spatial decomposition. The system is decomposed into several subsystems. Local optimization can be carried out for each subsystem. Then, the subsystems are synthesized to form the overall system. The first subsystem is composed of the pumping station that elevates the water from the supply source until the regulation reservoir. The rest of subsystems are composed of the possible booster pumps (or pumps in series) that are normally placed on a distribution line some distance from water source and serve to increase the pressure to downstream points on the line.
- Time decomposition. The operation of this optimization model is divided into two levels, long- and short-term models. The two models cover a planning period (annual season) and an operational period (a day), respectively. The annual period is divided into daily time steps and the daily period is divides into hourly time steps for the purpose of the optimization procedure.

(b) Water Demand

The model is based on the water hourly demands because the energy tariffs are based on time of day and because a regulation reservoir is usually designed to satisfy the fluctuating water demand during a day. The hourly demand patterns during the annual operation period can be calculated with historical data of the water hourly demand. When historical data of water demand are not available, simulation models are generally used (Pulido-Calvo *et al.*, 2006b).

(c) Long-Term Model

The purpose of the long-term model is to choose pump combinations and a storage capacity that allow to elevate the water in the hours with the most advantageous energy cost, to select the most appropriate electricity tariff and to establish an operation of pumping adapted with the time-of-use energy tariff while satisfying water demands and system requirements during the annual season. It is expected that the computational time for this model may be long. For this reason, the procedure starts with an initial solution for the selection of the pump combinations. The combinations that give the maximum requirements of flow and energy head of distribution network are chosen as initial solution and are used in the optimization algorithm of the design and operation of the pumping stations and the regulation reservoir (see section 4).

(d) Short-Term Model

The short-term model is used to update the solutions of the long-term model. This model is similar to the long-term model, except for the following item. The operational period must be short and appropriate for real-time operation. Typically, the operating plan is prepared for a period of twenty-four hours ahead, because the demands display a pronounced daily cycle, and energy tariffs are based on time of day. The real-time operation is based on the forecast of water demands. To accomplish this requirement, the historical data of the water demand must be considered. Methodologies for water demand forecasting can be time series approach, regression models or computational neural networks (Griñó, 1992; Jain *et al.*, 2001; Zhou *et al.*, 2002; Pulido-Calvo *et al.*, 2003; Bougadis *et al.*, 2005; Alvisi *et al.*, 2007; Pulido-Calvo *et al.*, 2007; Firat *et al.*, 2008). The results from the short-term model include which combination of pumps operates and their schedules for all the pump stations.

4. Initial Solution for the Selection of the Pump Combinations

Individual pump performance is expressed by energy head discharge, H-Q, and power consumption discharge, P-Q, curves. The H-Q and P-Q data points obtained from manufacturer's curves were entered for each pump size and type (range of pump discharge : 1-5000 m^3/h and range of pump head: 1-180 m). Generally, several pumps of similar sizes may operate in parallel to satisfy the different flow and head requirements at the pumping station. Second order polynomial equations for H-Q and P-Q were calculated for each pump type and for all possible pump combinations using least square regression techniques (Sabet and Helweg, 1989; Mays, 2000; Moradi-Jalal *et al.*, 2003; Pulido-Calvo *et al.*, 2003b). The polynomial coefficients were stored in data files.

Given the characteristics of the distribution system and the hourly water demands, the required energy head at each time period in the pumping to the water delivery is calculated applying the energy equation. The friction head losses are calculated applying the Darcy-Weisbach equation and the friction factor is estimated by means of the Colebrook-White equation (Mays, 2000).

Several pump combinations can satisfy the maximum requirements of flow and energy head of the water distribution system. The search procedure for selecting pump combinations divides the maximum flow in 2, 3, ..., n times. This way, for each value of division k ($2 \leq k \leq n$), are chosen the pumps that give the flow (Q/k) with energy head equal or higher than the maximum requirements at the water distribution network. If a pump combination can supply the flow but not at required energy head, a penalty is added to discourage selection of this combination. The pump combinations are also chosen with efficiency (required power/consumed power) equal or higher than 30% (large losses of power consumption are avoided). These pump combinations are only those that use in the optimization algorithm of the design and operation of the pumping stations and regulation reservoir.

5. Long-Term Model for the Optimal Design and Operation of Pumping Stations and Regulation Reservoir

The formulation of the optimal pump scheduling problem relies on the following set of assumptions:

(a) Pump Groups

Each pumping station includes one or more combinations of parallel pumps. A parallel pump combination can be a source or booster combination. A source combination pumps water from a borehole source into the origin reservoir. A booster combination pumps water out of an origin reservoir and into destination network distribution. The pumping station including a source combination is referred to as the source station. A pumping station including only booster combinations is referred to as a booster station.

(b) Regulation Reservoir Design

For high regulating capacities, the use of low cost materials such as soils or lands has been widely used in the construction of reservoirs for storing water. The cost of theses reservoirs has as relevant factors: the reservoir excavation, the surface to waterproof and the land surface where is the reservoir. The reservoir is assumed to be square (Pulido-Calvo *et al.*, 2006b). A reservoir cross section with relevant geometric parameters is given in Figure 1. As indicated, N1 is the outside levee slope (horizontal to vertical), N2 is the inside levee slope, Ta is the top width (m) of the levees, L is the length (m) of the base of the reservoir, H1 is the depth (m) of excavation below the original ground surface, H2 is the above-ground depth (m) of water storage, and F is the freeboard (m). The total storage volume VT (m^3) is given by:

$$VT = L^2\ (H1 + H2 + F) + 2\ L\ N2\ (H1 + H2 + F)^2 + 1.33\ N2^2\ (H1 + H2 + F)^3 \quad (4)$$

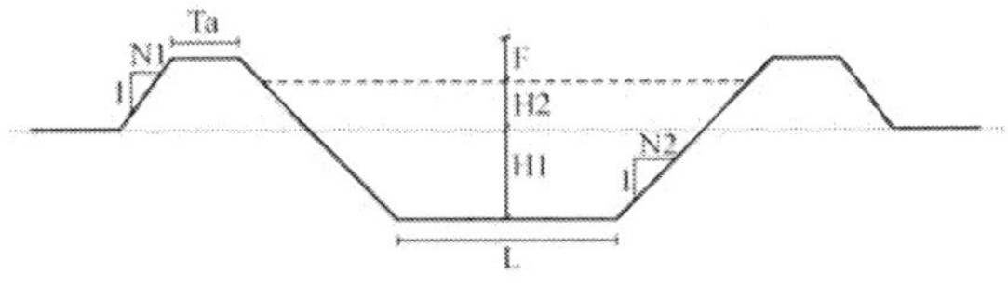

Figure 1. Reservoir cross-section and parameters

The cost of reservoir excavation is proportional to the excavated volume VX (m^3), which is computed as:

$$VX = L^2\ H1 + 2\ L\ N2\ H1^2 + 1.33\ N2^2\ H1^3 \quad (5)$$

It is desirable to minimize the excavation cost, and this occurs when VX is equal to the volume of fill used to construct the levees. The volume VL (m^3) comprised by the levees is calculated from:

$$VL = 4\left[L + 2\ N2\ (H1 + H2 + F) + 2\ Ta + 2\ N1\ (H2 + F)\right]\left[0.5\ (N1 + N2)\ (H2 + F)^2 + Ta\ (H2 + F)\right] \quad (6)$$

$$VL = VX\ (1 + CF) \quad (7)$$

where CF is the cut to fill ratio.

In practice, the quantities N1, N2, Ta, F and CF may be considered fixed with values dependent on local construction methods and guidance from technical assistance agencies. In this case, the values of N1 = 2, N2 = 3, Ta = 5 m, F = 1 m, and CF = 10% are considered. Due to safety, it is also common to construct reservoirs so that the total storage depth (H1 + H2) is in the range of 2 to 12 m. These factors are accounted for in model by the requirement that N1, N2, Ta, F and CF be fixed inputs. An iterative solution to calculate the dimensions H1, H2 and L for any capacity is required. The procedure starts with a trial value of (H1 + H2) and with the equation (4) is obtained L subject to L > (H1 + H2). The values of H1 are calculated with (7) using Newton numerical method. Then (H1 + H2) is incremented and new values of H1, H2 and L are computed. The iterative process stops when the capital cost of the reservoir is minimum.

(c) Reservoir operation. Operational Cost of the Pumps

An algorithm has been developed to determine the reservoir's optimal operation for a given regulating capacity. The algorithm is based on the concept of 'emptying period' (Pulido-Calvo *et al.*, 2006b). This is defined as the hourly interval ($i < j \leq i+k$) whereby at the initial hour t_{i+1} the reservoir has an useful stored volume and at the final hour t_{i+k} there is a deficit (when the reservoir is in deficit it is below the lower freeboard). There will be several emptying periods throughout the irrigation season. If V_j is the useful stored volume in hour j, the deficit volume VR_j is:

$$VR_j = V_{max} - V_j \quad (8)$$

where V_{max} is the maximum useful regulating volume. The initial value of V_i ($t = 0$) is V_{max}, that is, the algorithm is first applied when the reservoir is filled to capacity.

The 'potential hourly supply' (PHS) is defined as the volume supplied to the reservoir in one hour when pumping at the peak rate. The decision variable vector $\overline{E}(E_i,..,E_j,..,E_{i+k})$ represents the volumes pumped at each hour j. At each hour j, several pumpings can be carried out in different emptying periods (u,v,..,z): $E_j = E_{j,u} + E_{j,v} + ... + E_{j,z}$.

The reservoir deficit at the end of any emptying period v should be corrected by incrementing the volume of water stored at any hour j during that period ($i < j \leq i+k$). The algorithm will select the hour j when energy tariffs are lowest during the emptying period v. The volume to be pumped in hour j is conditioned by:

— The reservoir deficit at the end of the emptying period v, $-V_{i+k,v}$.
— The difference between potential hourly supply and the volume pumped at hour j during any emptying period u prior to v, $PHS - E_{j,u}$.
— The minimum deficit volume of the hours between hour j selected for pumping and the final hour (i+k) during emptying period v, min $(VR_{j,v}, VR_{j+1,v},..., VR_{i+k,v})$.

The volume to be incremented at hour *j* and emptying period *v* by pumping will be:

$$\Delta(V_{j,v}) = \min\left[-V_{i+k,v}, \min(VR_{j,v}, VR_{j+1,v}, ..., VR_{i+k,v}), PHS - E_{j,u}\right] \tag{9}$$

Once the volume is incremented at hour j, by incrementing $E_{j,v} = \Delta(V_{j,v})$, the available reservoir volumes V_j for the interval between hour j selected for pumping and the final hour (i+k) of the emptying period v will be incremented at an equal magnitude. And the deficit of the reservoir at the final hour (i+k) of emptying period v ($-V_{i+k,v}$) will be eliminated or corrected. One of the three following conditions is satisfied:

— The deficit at hour i+k ($-V_{i+k,v}$) is eliminated, thus meeting the demands of emptying period v [eq. (10)]. The following hour (i+k+1) is immediately analyzed and, in the case of deficit, will be corrected as described. The emptying period will then be equal to the previous emptying period and incremented by one hour: ($i < j \leq i+k+1$).

$$\Delta(V_{j,v}) = -V_{i+k,v} \tag{10}$$

— The deficit at hour i+k is not eliminated, but reduced. The new deficit is calculated by eq. (11), with an increase in volume given by eq. (12). As supply at hour j will be equal to the PHS, this hour cannot be used to correct the new deficit. Thus, the hour j must be reassigned to the emptying period v to correct the new deficit. This new hour will be the second hour with lowest energy tariffs during the emptying period v.

$$-V_{i+k,v} = -V_{i+k,v} + \Delta(V_{j,v}) \tag{11}$$

$$\Delta(V_{j,v}) = PHS - E_{j,u} \tag{12}$$

— The deficit at hour i+k has not been eliminated, but reduced [equation (11)], with an increase in volume given by eq. (13). The emptying period will have been reduced and the new period initiated at hour h with ($j < h \leq i+k$).

$$\Delta(V_{j,v}) = \min(VR_{j,v}, VR_{j+1,v}, \ldots, VR_{i+k,v}) \quad (13)$$

Eqs. (10)-(13) represent the possible cases after pumping in the hour j of emptying period v to satisfy a deficit $-V_{i+k,v}$. The same procedure is carried out for the new deficit and the new emptying period until the iterative process is completed for the entire operating period (irrigation season). Thus vector $\bar{E}$ will obtain the distribution of volumes pumped at each hour t during the irrigation season for any storage volume. $\bar{E}$ will determine the power consumption P for each hour t and hence the energy cost.

6. Short-Term Model for the Optimal Operation of Pumping Stations and Regulation Reservoir

This chapter evaluates the performance of linear multiple regressions and feed forward computational neural networks (CNNs) trained with the Levenberg–Marquardt algorithm (Sheperd, 1997) for the purpose of daily water demand modeling. The models are established using data recorded from the water distribution systems. The input or independent variables used in various CNN and multiple regression models are: (a) water demands from previous days; (b) water demands and climatic data (rainfall, maximum, minimum and average temperatures, relative humidity and wind speed) from previous days.

To assess the performance of the neural networks and the multiple regressions during the validation phase and therefore to identify the best short-term model, two measures of accuracy were applied: determination coefficient R^2 and efficiency coefficient E (Pulido-Calvo *et al.*, 2007).

7. Model Applications

The proposed methodology was applied to the demand pressurized system of the irrigation district of Fuente Palmera and to the water delivery of Hidrorecursos S.A., an intensive eel fishfarm, both located in the southern Spain. The purpose of these applications was to simulate the costs that would have been incurred if the pumping stations were designed according to model developed in this paper, and to compare these costs with the actual costs.

Input data for the model include a regulating reservoir and pumps useful life of 20 years, 5% interest rate and opportunity cost incurred by the loss of annual income (0.026 €/m^2) due to the alternative use of the area occupied by the reservoir. Other parameters that influence the cost of the reservoir include the cost of waterproofing material (2.40 €/m^2 for high-density, 1.5 mm thick polyethylene) and soil excavation (2.70 €/m^3).

The energy tariff times are: 8 off-peak hours (0.026 €/kWh in low electrical season – May, June, August and September– and 0.029 €/kWh in average electrical season –March, April, July and October–), 12 average hours (0.045 €/kWh in low electrical season and 0.050 €/kWh in average electrical season) and 4 peak hours (0.076 €/kWh in low electrical season and 0.085 €/kWh in average electrical season) per day. The off-peak hours are 12 p.m. to 8 a.m. and the peak hours are 10 a.m. to 2 p.m. All other hours are considered average energy tariff.

7.1. Water Distribution System of an Irrigation District

The demand pressurized system of the irrigation district of Fuente Palmera, located in the Guadalquivir valley (southern Spain), has an annual water consumption of 16.5 ± 5.9 hm^3 and must be drawn from the Guadalquivir River. The average irrigated area is approximately 5,000 ha and is irrigated by sprinkling on demand.

Nowadays the pressurized irrigation system has two pumping stations in series (Figure 2). The first station (4 pumps in parallel; each pump group has an electric motor that operates at 990 rpm and the power motor is 2,500 hp) carries water from the Guadalquivir River to a 5000 m^3 tank, that is the aspiration chamber for the second station (6 pumps in parallel; each pump group has an electric motor that operates at 990 rpm and the power motor is 2,500 hp) which discharges directly into the distribution line. Given that the storage capacity of this tank does not allow the two pumping stations to operate independently, it is used to provide pressure to the branched pipeline system.

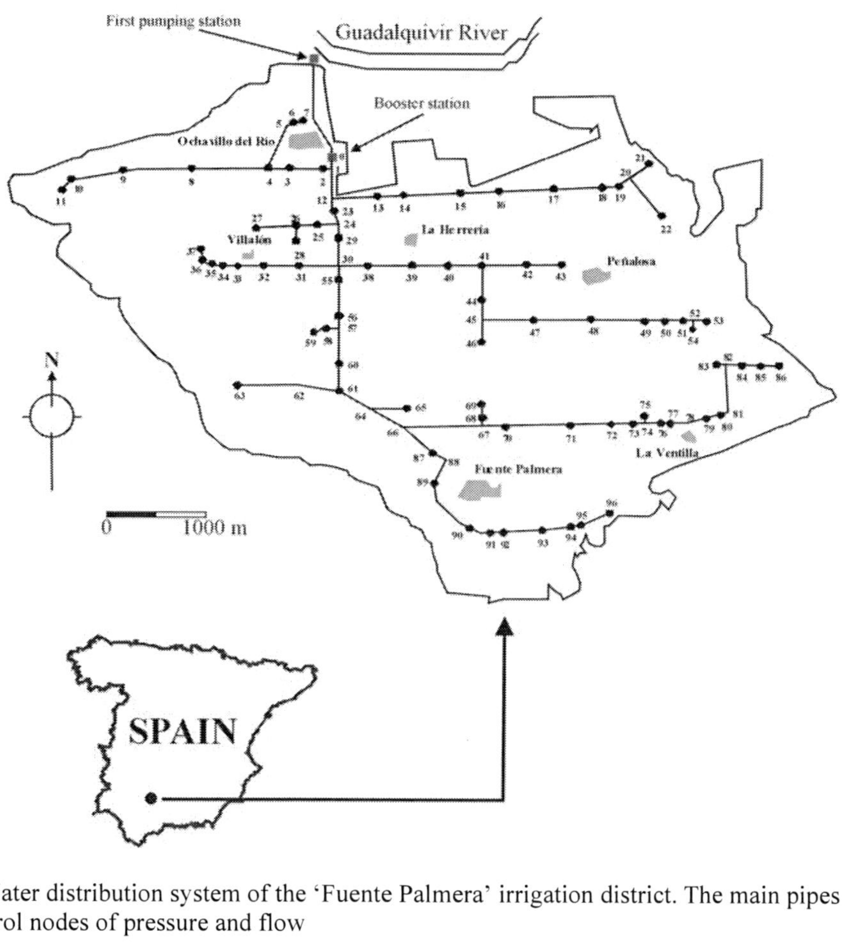

Figure 2. Water distribution system of the 'Fuente Palmera' irrigation district. The main pipes network has 96 control nodes of pressure and flow

The main water supply system carries water from the booster station (second pumping station) to 78 different groups of farmers, each one whom has only one outlet. The minimum, maximum, and average areas of the group of farmers are 21.6, 218.3, and 67.4 ha, respectively. From the main network outlets, the water is distributed to the plots through a secondary pipe network that is underground and fixed. The average area of the plots is 6.25 ha. Nine of the more representative crops were selected according areas occupied in a period of 14 consecutive irrigation seasons (from 1984-1985 to 1997-1998): cotton, sunflower, wheat, sugar beet, olive, corn, sorghum, citric fruits and melon/watermelon. The water requirements of this irrigation district are shown in Figure 3.

7.2. Water Distribution System of a Fishfarm

In Hidrorecursos S.A., an intensive eel fishfarm located in the province of Córdoba (southern Spain), the water is pumped from the Puente Nuevo reservoir (Figure 4). The main pumping station has a pump combination with two groups in parallel. Each pump group has an electric motor that operates at 1,480 revolutions per minute (rpm) and the motor power is 300 horsepower (hp) (220.8 kW). In general, the pump operation scheme is at a constant rate and the excess flow is dumped when the water demand is lower.

In this intensive fishfarm, the water is pumped from the Puente Nuevo reservoir to a main channel with slope of 0.1%. The water is then transported by this channel by gravity to seven tanks series that have the spatial configuration shown in Figure 4. The 'A', 'B', 'C' and 'D' series are the nursery tanks (eel weight: 0.3-40 g) that have 12 circular tanks with a capacity of 3.2 m^3 in each series (12 × 3.2 m^3 tanks). The 'E' and 'F' series are the pre-growth tanks (eel weight: 40-110 g) with 8 rectangular tanks of 16 m^3 (8 × 16 m^3 tanks) and 2 rectangular tanks of 32 m^3 (2 × 32 m^3 tanks), respectively. In the last tank 'G' series (14 × 110 m^3 rectangular tanks), the eels grow to commercial weight (150 g). The water requirements of this fishfarm during the years 1999 and 2000 are shown in Figure 5.

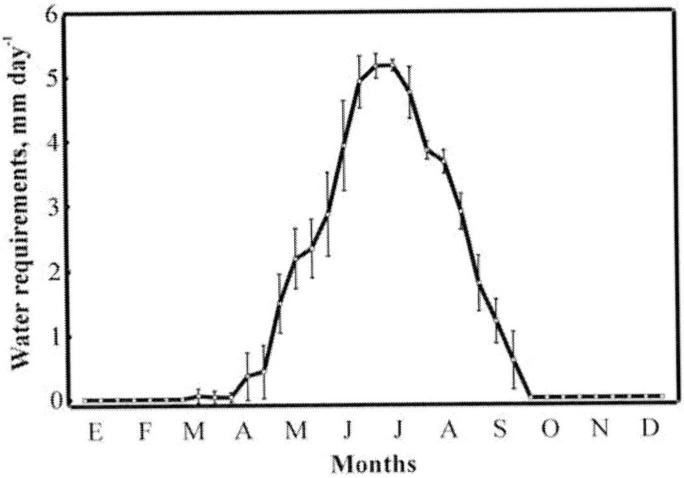

Figure 3. Water requirements of the 'Fuente Palmera' irrigation district

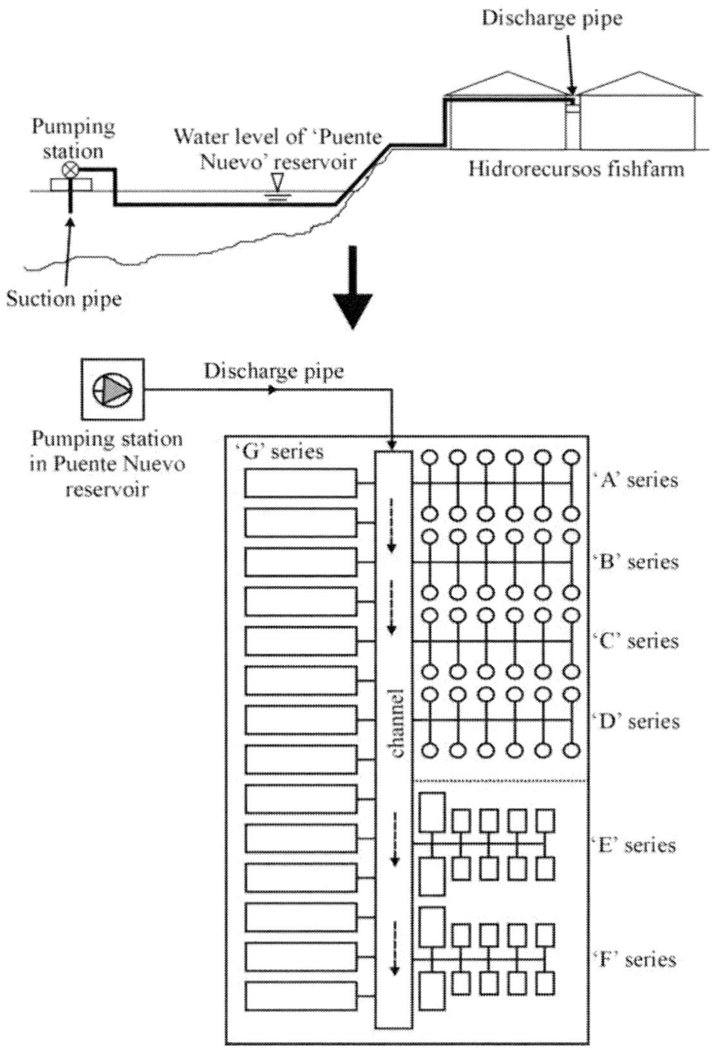

Figure 4. Scheme of the water distribution network of the 'Hidrorecursos' fishfarm

7.3. Results: Irrigation District

The least total costs were obtained with the alternative that considers a regulating reservoir between the two pumping stations with the following characteristics: (a) source pumping station with 8 groups in parallel with electric motors of 340 hp/pump and 1,480 rpm; (b) booster station with 7 groups in parallel with electric motors of 340 hp/pump and 1,480 rpm; (c) useful capacity for storing water is determined to be 65,000 m^3, which is 41% of the maximum daily demand (158,000 m^3/day) and 0.42% of total demand (15,400,000 m^3) for the irrigation district. A lower and upper freeboard is added to this volume, thus obtaining a total volume of 91,000 m^3 for the regulating reservoir. The dimensions L, H1 and H2 (Figure 1) are 48.63 m, 5.32 m and 6.68 m, respectively.

Figure 5. Water demand of the 'Hidrorecursos' fishfarm

Table 1. Costs (€/year) of water supply system with and without regulating reservoir

	Without reservoir (current situation of irrigation district)	With reservoir (optimal solution)
First pumping station (operation cost) (€/year)	407,000	153,000
Second pumping station (operation cost) (€/year)	513,000	304,000
Total cost (operation cost + amortized capital cost) (€/year)	1,070,000	633,000

With regard to energy consumption, the cost of the first pumping station with the regulating reservoir (153,000 €/year) is 62% less than the cost of pumping directly to the water supply system (407,000 €/year) (current performance of the pumping station that discharge the water demand of each hour). Also, the cost of the second pumping station in the optimal solution (304,000 €/year) is 41% less than the cost of pumping directly to the water supply system (513,000 €/year). The annual total cost, which includes the storage construction cost and the cost of operation and capital of the pumps, is 41% less with the regulating reservoir and investment is amortized in two years time (Table 1). Thus, energy costs are considerably reduced when a reservoir is used to adapt pumping hours to time-of-use energy tariffs. This results in a decrease in average and peak energy use and an increase off-peak energy use, and greater pump efficiencies.

The optimal operating scheme for the first pumping station with the reservoir demonstrates that only off-peak energy tariff times should be used at the start of irrigation. As water demands increase, the use of off-peak hours will be incremented until the peak rate is pumped during all of the 8 off-peak hours (June 2nd). This operating scheme will be maintained until average energy tariff times are needed (June 6th) to meet demand. The use of average hours also increases during periods of maximum water demand (June 21st-July 27th) although all the hours will not be used and potential hourly supply will not be pumped. Pumping at these times subsequently decreases, becoming null at the end of August (August 20th) when water will only be pumped to the reservoir during off-peak hours. It is not necessary to pump water to the regulating reservoir during peak energy tariff times.

The frequency distribution of pump efficiencies and regulating efficiencies for the first pumping station during the entire season with and without regulating reservoir were compared. Mean pump efficiency ($\overline{\eta}$) and mean pump regulating efficiency ($\overline{\eta}_{reg}$) for the first pumping station are 79.91% and 92.82%, respectively, with a regulating reservoir, and 77.22% and 85.49% without one. When comparing the frequency distributions of both situations (with and without regulating reservoir) by means of the χ^2 test, significant differences are obtained in the distributions of pump efficiency ($\chi^2 = 685.81$; $p < 0.001$) and in the distributions of pump regulating efficiency ($\chi^2 = 2131.82$; $p < 0.001$).

7.4. Results: Fishfarm

The optimum alternative doesn't imply the inclusion of regulating reservoir. The pump combination has nine groups in parallel. Each pump group has an electric motor that operates at 2,900 revolutions per minute (rpm) and the motor power is 40 horsepower (29.4 kW). The operator will have to turn on and off the pumps during the annual operation period in accordance with the water requirements. It is clear, by using this optimization model, a decrease of about 92% is obtained in annual operation cost and annual depreciation cost of the initial investment compared to actual situation (average total annual cost of fishfarm pumping station = 72,000 €/year; optimal solution = 6,210 €/year).

Given that the water requirements of the fishfarm are different over the course of a annual operation period, the optimum pump combination is the alternative that operates more closely adapted to system head curve, that is to say, the pump combination with the highest regulating efficiencies during all the annual operation period.

The results showed in sections 7.3 and 7.4 were achieved through the use of the aforementioned inputs as well as the optimization equations implemented in DYGOSIA v.1.0 computer program (Figure 6).

7.5. Water Demand Forecasting for Short-Term Model

For both the neural models and multiple regressions the best values for the evaluation magnitudes R^2 y E were obtained when water demand for the two days prior to forecasting was used as the input or for the independent variables. In all of the models considered, the neural networks provide better forecasts than the multiple regressions. In almost all the neural models, values for R^2 were higher than 0.7 and for E were lower than 30%.

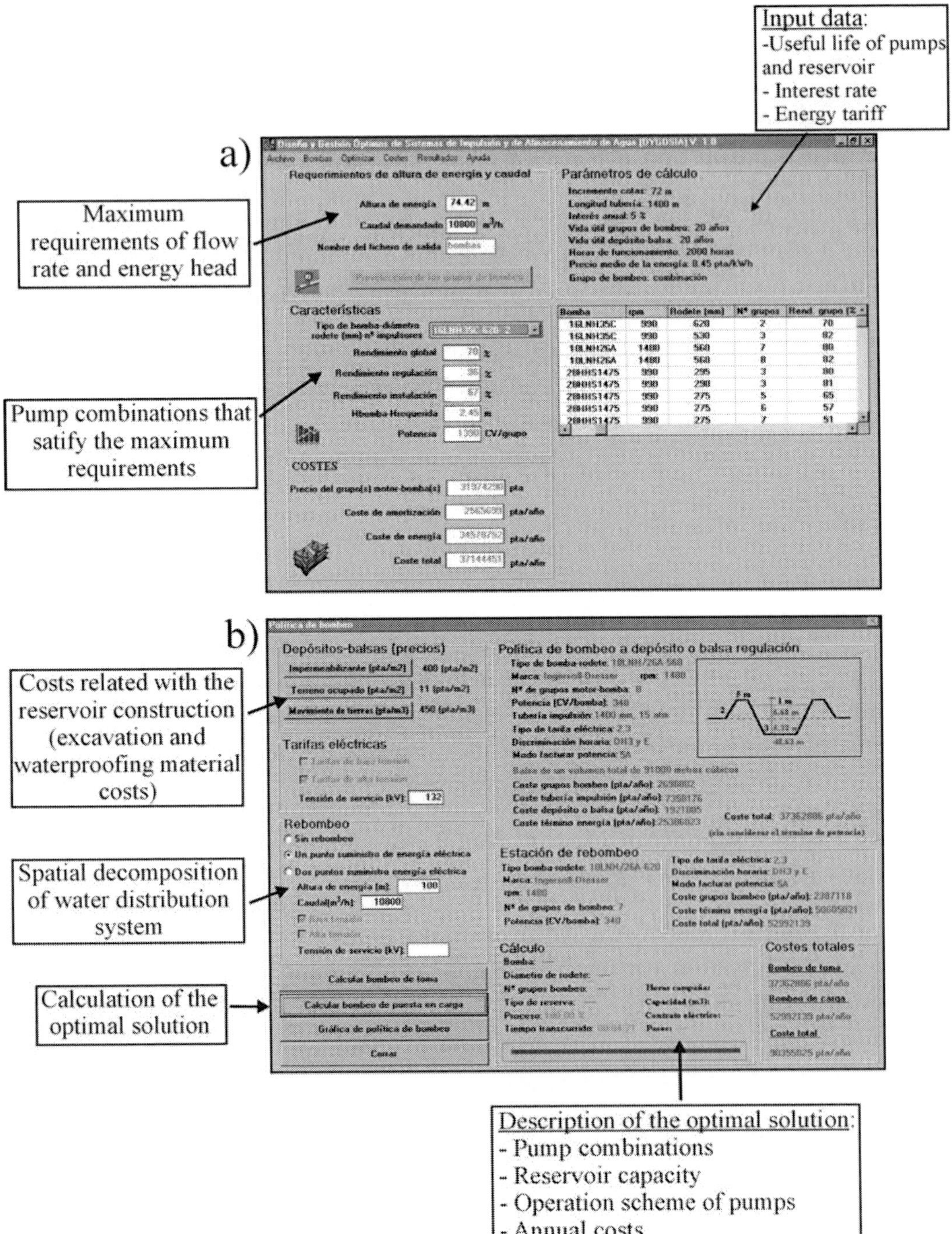

Figure 6. Main windows of DYGOSIA v.1.0 computer program (spanish version)z

8. CONCLUSION

The energy required for operating pumping stations in water distribution systems may be significant. Thus, an optimization model has been developed to identify the pump, or pump combination, and reservoir storage capacity which should be working in order to satisfy the water demand at minimum annual total cost (annual operation cost plus annual depreciation cost of the initial investment). The difficulty of the discrete pump discharges have been considered in this model.

To support operational decisions, a computer software package for the model was also coded and applied successfully to two real delivery systems. It was found that the proposed model may significantly reduce the total annual costs.

The model was developed so that it should be applicable for other water supply system by introducing a similar set of data. It will be very effective if the water daily demand simulation is accurate. This approach of water daily requirements is highly related with the available information/data of the facilities. This way, in those water supply systems with a high control level of their operation schemes will allow to carry out a more easy and accurate estimation of water daily demands.

The addition of a regulating reservoir is feasible if a reduction in the total cost of providing water is achieved. The developed economic analysis indicated that the addition of a regulating reservoir with optimal size and operation scheme is cost effective in the irrigation district but not in the fishfarm. The optimal solution with regulating reservoir saved 41% of the annual total cost in the irrigation district. The optimal solution without regulating reservoir saved 92% of the annual total cost in the fishfarm.

Given that agreement is necessary between the desired height of the structural sections, optimal storage capacity, total volume of extracted soil, and the balance between extracted material and the material used in the dyke, the storage capacity is determined by the landform of the construction site. A trapezoidal cross-section reservoir with the square base has been used in the model developed in this paper. This type of reservoir may not be used if the terrain is not suitable. However, the model developed herein can be considered an initial approach to the design of storage facilities under these circumstances.

However, it should be noted that the provision of a regulating reservoir is not always the most suitable solution, as energy savings may not warrant the initial investment. This may be the case, for example, when the reservoir is located too far from the benefited area or the pumping station, resulting in friction head losses in the pipelines which would necessitate greater energy requirements and thereby increase total cost. To obtain these results it would be necessary to perform a comparative cost study with concrete data from the water supply system under study.

In this chapter, consumer water demand forecasting systems that can support decision making of the water delivery administrators are proposed using multiple regressions and computational neural networks. Determination coefficients higher to 92% and efficiency coefficient E higher to 0.91 have been obtained in the validation period, when water demand of the two days prior to forecasting was used as input or independent variables to the neural network. The neural models performed better than the regressions. Short-term demand modeling can be used as input in methods and/or programs for the management of water delivery systems in real time. Furthermore, this approach achieves a better fit of the pumped volumes and the real demand of the distribution network, thereby leading to a more rational use of water and energy resources.

REFERENCES

Alvisi, S., Franchini, M. & Marinelli, A. (2007). A short-term, pattern-based model for water-demand forecasting. *J. Hydroinformatics*, *9(1),* 39-50.

Biscos, C., Mulholland, M., Le Lann, M. V., Buckley, C. A. & Brouckaert, C.J. (2003). Optimal operation of water distribution networks by predictive control using MINLP. *Water SA*, *29(4),* 393-403.

Bougadis, J., Adamowski, K. & Diduch, R. (2005). Short-term municipal water demand forecasting. *Hydrol. Process.*, *19,* 137-148.

Breytenbach, P., Meiring, J. A. & Oosthuizen, L. K. (1996). The importance of the cost of electricity in irrigation farming. *Water SA*, *22(4),* 333-338.

Brion, M. & Mays, L. W. (1991). Methodology for optimal operation of pumping stations in water distribution systems. *J. Hydraul. Eng.*, *117(11)*, 1-19.

Buchleiter, G. W. & Heermann, D. F. (1986). Using computers to manage irrigation systems. *J. Water Resour. Plan. Manage.*, *112(3),* 354-365.

Buchleiter, G. W., Heermann, D. F. & Duke, H. R. (1984). Integrated water-energy management system for center pivot irrigation: functional requirements. *Trans. ASAE*, *27,* 1419-1423.

Buchleiter, G. W. & Heermann, D. F. (1990). Management of multiple pump stations. *Applied Eng. Agric.*, *6(1)*, 39-44.

Cembrano, G., Wells, G., Quevedo, J.; Pérez, R. & Argelaguet, R. (2000). Optimal control of a water distribution network in a supervisory control system. *Control Eng. Pract.*, *8*, 1177-1188.

Crawley, P. D. & Dandy, G. C. (1993). Optimal operation of multiple-reservoir system. *J. Water Resour. Plan. Manage.*, *119*, 1-17.

Duke, H. R., Heermann, D. F. & Blue, M. C. (1984). Computer control of irrigation for electrical load management. *Trans. ASAE*, *27,* 597-602 & 608.

Firat, M., Yurdusev, M. A. & Turan, M. E. (2008). Evaluation of artificial neural network techniques for municipal water consumption modeling. *Water Resour. Manage.*, DOI 10.1007/s11269-008-9291-3.

Griñó, R. (1992). Neural networks for univariate time series forecasting and their application to water demand prediction. *Neural Network World*, *2(5),* 437-450.

Heermann, D. F., Buchleiter, G. W. & Duke, H. R. (1984). Integrated water-energy management system for center pivot irrigation: implementation. *Trans. ASAE*, *27*, 1424-1429.

Hirose, S. (1997). Determination of the capacity of a regulating pond in a pipeline irrigation system. *Rural Env. Eng.*, *33*, 66-78.

Jain, A., Varshney, A. K. & Joshi, U. C. (2001). Short-term water demand forecast modelling at IIT Kanpur using artificial neural networks. *Water Resour. Manage.*, *15(5),* 299-321.

Joalland, G. & Cohen, G. (1980). Optimal control of water distribution network by two multilevel methods. *Automatica*, *16*, 83-88.

Jowitt, P. W. & Germanopoulos, G. (1992). Optimal pump scheduling in water-supply networks. *J. Water Resour. Plan. Manage., 118(4),* 406-422.

Kerr, N. M. (1981). Design of equipment and selection of materials-an engineer's assessment. In: Tiews, K. (Ed.), *Aquaculture in Heated Effluents and Recicurlation Systems*, Vol. I. Heenemann Verlagsgesellschaft mbH, *Berlín*, pp. 151-181.

Lansey, K. E. & Mays, L. W. (1989). Optimization model for water distribution system design. *J. Hydraul. Eng., 115(10)*, 1401-1418.

León, C., Martín, S., Elena, J. M. & Luque, J. (2000). EXPLORE-Hybrid expert system for water networks management. *J. Water Resour. Plan. Manage., 126(2)*, 65-74.

Mays, L. W. (2000). *Water Distribution Systems Handbook.* McGraw-Hill Companies, Inc., New York.

Mehta, B. K. & Goto, A. (1992). Design and operation of on-farm irrigation ponds. *J. Irrig. Drain. Eng., 118(5),* 659-673.

Moradi-Jalal, M., Mariño, M.A. & Afshar, A. (2003). Optimal design and operation of irrigation pumping stations. *J. Irrig. Drain. Eng., 129(3)*, 149-154.

Nel, D. & Haarhoff, J. (1996). Sizing municipal water storage tanks with the Monte Carlo simulation. *J. Water SRT-Aqua, 45(4),* 203-212.

Nitivattananon, V., Sadowski, E. C. & Quimpo, R. G. (1996). Optimization of water supply system operation. *J. Water Resour. Plan. Manage., 122(5),* 374-384.

Ormsbee, L. E., Walski, T. M., Chase, D. V. & Sharp, W. W. (1989). Methodology for improving pump operation efficiency. *J. Water Resour. Plan. Manage., 115,* 148-164.

Planells, P., Carrión, P., Ortega, J.F., Moreno, M.A. & Tarjuelo, J.M. (2005). Pumping selection and regulation for water-distribution networks. *J. Irrig. Drain. Eng., 131(3),* 273-281.

Pulido-Calvo, I., Roldán, J., López-Luque, R. & Gutiérrez-Estrada, J.C. (2003a). Demand forecasting for irrigation water distribution systems. *J. Irrig. Drain. Eng., 129(6),* 422-431.

Pulido-Calvo, I., Roldán, J., López-Luque, R. & Gutiérrez-Estrada, J.C. (2003b). Water delivery system planning considering irrigation simultaneity. *J. Irrig. Drain. Eng., 129(4),* 247-255.

Pulido-Calvo, I., Gutiérrez-Estrada, J. C. & Asensio-Fernández, R. (2006a). Optimal design of pumping stations of inland intensive fishfarms. *Aquacul. Eng., 35,* 283-291.

Pulido-Calvo, I., Gutiérrez-Estrada, J. C., López-Luque, R. & Roldán, J. (2006b). Regulating reservoirs in pressurized irrigation water supply systems. *J. Water SRT-Aqua, 55(5),* 367-381.

Pulido-Calvo, I., Montesinos, P., Roldán, J. & Ruiz-Navarro, F. (2007). Linear regressions and neural approaches to water demand forecasting in irrigation districts with telemetry systems. *Biosystems Eng., 97,* 283-293.

Pulido-Calvo, I., Gutiérrez-Estrada, J. C. & Corbacho, J. M. (2008). Pipes size selection of water distribution systems of fishfarms. *Aquacul. Eng., 39,* 43-52.

Quimpo, R. G. & Shamsi, U. M. (1991). Reliability-based distribution system maintenance. *J. Water Resour. Plan. Manage., 17,* 321-339.

Sabet, M. H. & Helweg, O. J. (1985). Cost effective operation of urban water supply system using dynamic programming. *Water Resour. Bull.*, 21, 75-81.

Sabet, M. H. & Helweg, O. J. (1989). Peaking storage tanks for water supply systems. *Water Resour. Bull., 25(5),* 1047-1053.

Shepherd, A. & Ortolano L. (1996). Water-supply system operations: Critiquing expert-system approach. *J. Water Resour. Plan. Manage.*, *122,* 348-355.

Shepherd, A. J. (1997). *Second-Order Methods for Neural Networks*. Springer, New York

Stetson, L. E., Watts, D. G., Corey, F. C. & Nelson I. D. (1975). Irrigation system management for reducing peak electrical demands. *Trans. ASAE, 18(2)*, 303-306 & 311.

Tarquin, A. J. & Dowdy, J. (1989). Optimal pump operation in water distribution. *J. Hydraul. Eng.*, *115(2)*, 158-168.

Walski, T. M. (1984). *Analysis of Water Distribution Systems*. Van Nostrand Reinhold, New York.

Zessler, U. & Shamir, U. (1989). Optimal operation of water distribution systems. *J. Water Resour. Plan. Manage.*, *115(6)*, 735-752.

Zhou, S. L., McMahon, T. A., Walton, A. & Lewis, J. (2002). Forecasting operational demand for an urban water supply zone. *J. Hydrol.*, *259,* 189-202.

In: Energy Costs, International Developments…
Ed: Leszek Kowalczyk and Jakub Piotrowski ISBN: 978-1-60741-700-2

Chapter 7

ADSORPTION MECHANISM OF METHANE AND ITS EFFECT ON SEPARATION AND STORAGE METHODS

Li Zhou*

High Pressure Adsorption Laboratory, School of Chemical Engineering and Technology, Tianjin University, Tianjin 300072, China

ABSTRACT

A brief review was given on the adsorption mechanism of methane. The effect of adsorption mechanism on separation and storage was interpreted. Methane as a supercritical gas adsorbs on solid surface merely monolayerly, and any novel material cannot assume the duty of methane/natural gas storage basing on adsorption. Instead, another mechanism was proposed for storage enhancement. Commentary was made on the research of methane separation with respects to the enrichment of methane or removal of carbon dioxide based on the adsorption mechanism.

INTRODUCTION

It is well known that methane has a much stronger greenhouse effect than carbon dioxide, and a huge amount of methane is being emitted into the atmosphere via different passages. Some emission passages are out of human control, but the others, for example, those escaped from coal mining, can be purposely controlled. The importance of methane is not only due to the greenhouse effect, but also due to the role in alternative fuels. Alternative fuel was put forward in 70's of last century, but it becomes urgently required right now due to the high price of crude oil and the serious situation of environment. Hydrogen energy is certainly the best solution for the problems we are facing in energy and environment. However, there are too many and tough problems in front of hydrogen energy to have it an available enrgy presently. On the other hand, methane-based fuel is environment benign and more or less sustainable. The major methane-based fuel is natural gas, which is not a sustainable source.

* Correspondence author: http://www.hpal-tju.com, E-mail: zhouli@tju.edu.cn

However, almost same amount of methane-based fuel contained in coalmine gas has not been exploited. The methane buried in flammable ice is estimated to range in 35,000 to 4,200,000 trillion cubic feet. The global volume may exceed the worldwide store of conventional fossil fuels [1], though unsuitable for exploitation presently. Firedamp is a sustainable source of methane. Firedamp is produced from agricultural residue and dejections serving as energy source in some countryside places of China, and it will be produced in large scale from urban garbage. This kind of methane-based fuel is sustainable. However, utilization of firedamp and coalmine methane including natural gas, needs an efficient technology of separation or enrichment of methane from its mixtures. An efficient storage technology has to be worked out in order for the replacement of liquid fuel with methane. Therefore, a lot of efforts were dedicated to separation and storage of methane since last 70's, and innumerous publications emerged since then. However, the state of the art is still not satisfactory from the viewpoint of industry, but there is a great potential for the utilization of methane-based fuel.

It is clear, separation and storage are kernel research topics for the utilization of methane based fuel. Methane is an incondensable gas at ambient condition; therefore, adsorption plays an important role in both separation and storage. Understanding the adsorption mechanism is very important because the mechanism functions in all where adsorption occurs; therefore, it is the mechanism that determines the efficiency/capacity of a separation/storage method. It is worthwhile to give a brief review on the theoretical description of methane adsorption and see how theoretical doctrine affects the research of methane separation and storage.

Adsorption Mechanism of Methane

Six types of adsorption isotherms were observed till presently, from which adsorption mechanisms of mono-molecular layer surface coverage, multi-layer coverage, volume-filling and capillary condensation were identified [2]. Two common features are shown on these isotherms. First, the amount adsorbed is a unimodal increasing function of adsorption pressure; second, saturation pressure is the upper bound of adsorption, beyond which condensation occurs. Addressing this kind of isotherms the wellknown adsorption theories were proposed, eg. the Langmuir equation, the BET equation, and the D-A equations [2], etc., basing on which porous solids have been characterized so far. However, there is another kind of adsorption isotherms. It features in showing a maximum and no limit for adsorption pressure was assumed. This is what observed on the adsorption of incondensable gases. Any gas is incondensable at temperatures higher than the critical temperature of the gas. Because fluids cannot be liquid at above-critical temperatures, saturation pressure doesn't have meaning, which is why the experimental pressure was reported as high as several hundreds MPa [3]. Adsorption of the incondensable gases is, therefore, synonymous as supercritical adsorption or high pressure adsorption. It didn't receive much attention until last 70's when alternative fuels became a hot topic of research. Natural gas and hydrogen were considered alternative of petroleum products. On-board storage of gaseous fuels is the first problem to solve, and the storage based on adsorption attracted world's attention since then. Methane, as the major component of natural gas, was extensively studied for the adsorption behavior on miscellaneous adsorbents. Hydrogen has a similar experience also. Either methane or hydrogen has a very low critical temperature (-82.6 and –240 °C, respectively). They are incondensable gases at ambient temperatures; therefore, how to explain the experimentally measured adsorption isotherms of methane and/or hydrogen logically became a topic of research. Should isotherms were reasonably modeled/explained with a theory, the theory would be able to guidance the research on separation or storage.

In the early stage of the study on supercritical adsorption, researchers were tried to explain/model the adsorption isotherms of incondensable gases by a known equation with an assumption of quasi-liquid or overheated liquid for the adsorbed phase, and different so-called empirical equations were proposed for the fictitious saturation pressure. However, such models can fit the experimental isotherms only for the section before maximum due to the unimodel property. Later on, the Ono-Kondo equation was used to describe isotherms with maximum and partial success was achieved [4]. However, it was shown that parameters in the equation didn't have physical meaning [5]. More pertinent fitting was achieved with multi-parameter equations of state [6] or that based on density function theory [7]. However, about eight or more parameters contained in these models; no physical meaning was clearly defined for these parameters; therefore, such models cannot provide any meaningful guidance for the interested research.

Appropriate explanation/modeling of supercritical adsorption isotherms was found in the classical definition of adsorption given by Gibbs [8,9]:

$$n = n^S - \rho_g V_a = V_a\left(\rho_a - \rho_g\right)$$

where V_a is the volume of adsorbed phase defined by a steep density gradient at the interface; ρ_g and ρ_a are the density of ambient gas and adsorbed phase, respectively. This definition indicates that adsorption, n, is a density-excess quantity. The so-called absolute adsorption, n^s, equals $\rho_a V_a$; is the total mass confined in the adsorbed phase. The difference between the absolute and the excess quantities of adsorption is the product $\rho_g V_a$, the mass of adsorptive confined in the "volume of adsorbed phase" without adsorption. This difference is not discussed for the adsorption of condensable gases because its value is numerically negligible in comparison with $\rho_a V_a$ due to the fact that $\rho_a >> \rho_g$. However, the gas density may be so large in the adsorption of supercritical gases that zero or negative value of n may experimentally be obtained [10]. Therefore, evaluation of the difference between n and n^s is necessary for supercritical adsorption, which is why the determination of n^s was considered the essential problem [11] or a major challenge [12] for the theoretical study of supercritical adsorption.

Based on the fact that the difference between n and n^s can be neglect for an adsorbed phase that has not been fully developed, the author proposed a strategy to establish the model of absolute adsorption relying on experimental isotherms [13,14]. According to the theory of adsorption potential, isotherms for different temperatures fall in a general isotherm [15]. Modeling the general isotherm with a Langmuir type equation, the data showing remarkable deviation from the feature of absolute adsorption, i.e. unimodal function, were skipped. Next, the Langmuir type model was expanded into Taylor series and the nonlinear terms were skipped. The remained data fall in the Henry law range; therefore, may stand for absolute adsorption. The linear model yields the data of absolute adsorption for the whole range tested. As consequence, the difference term in the Gibbs formula was evaluated. In so doing, all experimental isotherms have been satisfactory modeled basing on the classical Gibbs formula, no matter what supercritical gases and what adsorbents were tested as shown for example in

Figures 1 and 2 [16,17]. The isotherm equation applied in the Gibbs formula contains only three parameters with definite physical meaning [14]. Important conclusion regarding the adsorption mechanism of supercritical gases can be drawn due to the perfect agreement between model and experimental data. Based on the assumption of monolayer surface coverage, the intermolecular distance in the adsorbed phase was calculated. As shown in Figure 3, the intermolecular distance was a linear function of temperature, and the two plots standing for both phases meet at 50 K and indicate the distance of 0.34 nm, which is the distance between carbon atoms in graphite. Therefore, the assumption of monolayer surface coverage is correct. Applying the model to the experimental adsorption isotherms of CO_2 on crossing the critical temperature, perfect fit was also observed as shown in Figure 4 [18]. The number of molecular layer adsorbed on the surface of adsorbent was calculated from the adsorbed phase volume. As shown in Table 1, the number of molecular layer changed from more than one to less than one on crossing the critical temperature. This change was also observed in a FT-IR in situ measurement [19].

On developing a method or a material targeting at methane separation or storage, one must keep in mind that methane, as other supercritical gases, adsorbs on any sort of adsorbents only monolayerly. Importance of the adsorption mechanism will further be illustrated in the following sections.

Table 1. Adsorbate layers in the adsorbed phase.

T, K	307	313	318	323	340	360
Number of molecular layers	1.20	1.12	1.06	1.00	0.87	0.82

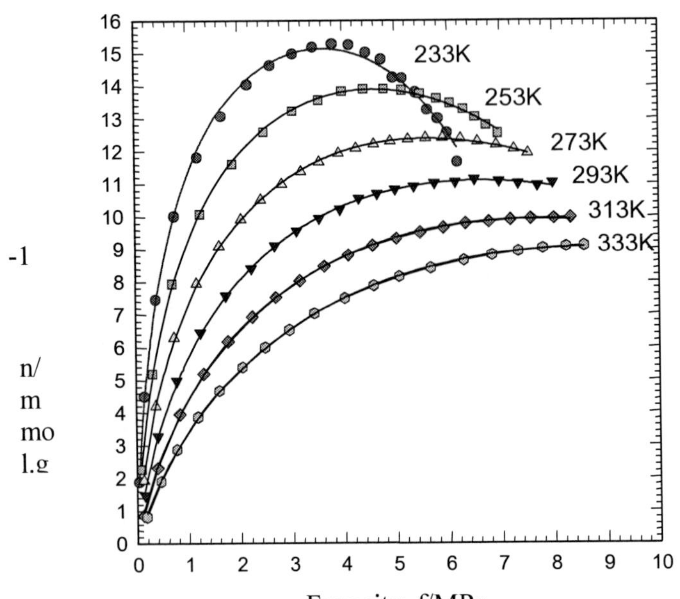

Figure 1. Adsorption isotherms of methane on activated carbon. Point: experimental data; Curves: Model generated [16].

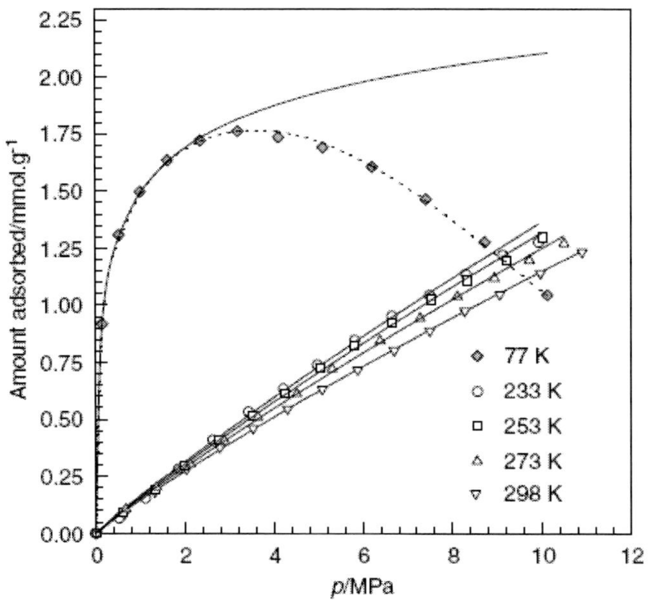

Figure 2. The excess and the absolute adsorption isotherms of H_2 on MWNT. Points: the excess data; solid curves: the absolute; dotted line: the model of the excess isotherm at 77 K [17].

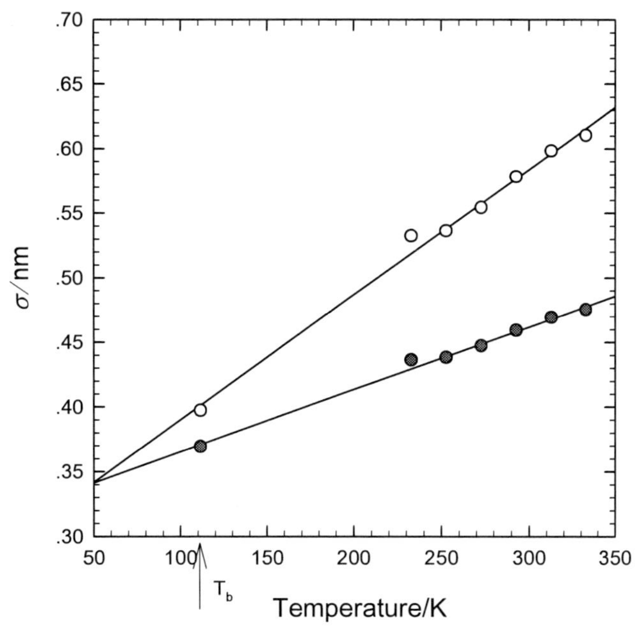

Figure 3. A Comparison of intermolecular distances. ○ : in free state; • : in the adsorbed state [16].

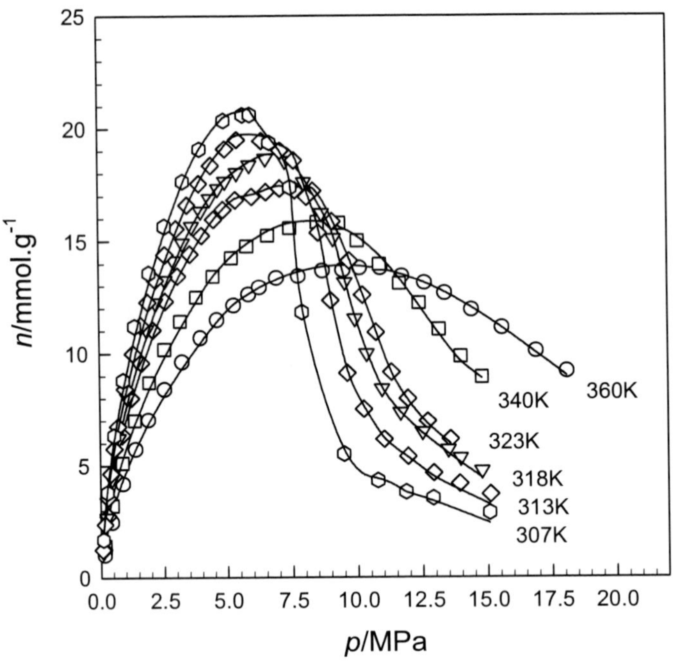

Figure 4. Adsorption isotherms of CO_2 for the supercritical region. Dots: experimental; Curves: predicted by model [18].

Methane Separation

The content of methane is less than 30 % in quite many coalmine gases. It is hopefully to enrich the methane up to 80 % with a low-cost technique. Major components of coalmine gas are methane and air. The adsorption amount of oxygen is very close to that of nitrogen; therefore, nitrogen can be a representive of key components for the separation. Separation between CH_4 and N_2 was considered one of challenges the researchers faced [20]. Many efforts were made to enlarge the separation coefficient between CH_4 and N_2. However, no much improvement was achieved. Compared to nitrogen, methane is preferably adsorbed on most materials, but the difference is not much. The separation coefficient reported in literature for different zeolites and activated carbons ranged in 2 to 3.6 [21,22]. Therefore, although the PSA (pressure swing adsorption) technology was tested and realized for the separation, the separation cost is high due to the low separation coefficient. Based on the monolayer adsorption mechanism of methane, the difference in the adsorption amounts between methane and nitrogen must be enlarged by increasing the specific surface area of adsorbents, and this was shown true as illustrated in Figure 5 [23]. Although local deviation was observed, a linear correlation was definitely shown. The adsorbents covered on the plot include activated carbons, zeolites and silica gel. This plot provides one more direct proof for the monolayer adsorption mechanism of methane. Better separation performance was achieved by using adsorbents of high surface area.

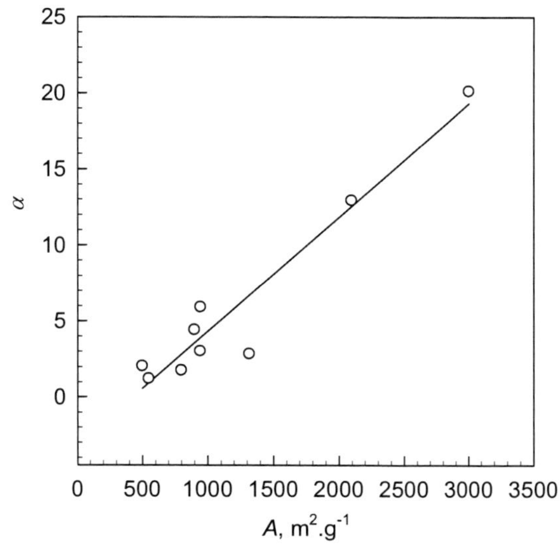

Figure 5. Correlation of separation coefficient between CH_4/N_2 with specific surface area of adsorbents [23]. .

Table 2. The adsorption amount of CO_2 and CH_4 on activated carbons at 0.4 MPa and 298 K.

	Surface area, m^2/g	n_{CO_2} mmol/g	n_{CH_4} mmol/g	n_{CO_2}/n_{CH_4}
AC-CS[1]	2587	5.8	2.9	2.0
AC-BB[2]	1567	3.8	1.3	2.9

Because most adsorbents adsorb methane more preferably than nitrogen, methane is not the column-top product, but obtained on regenerating the adsorbent. The adsorption pressure lost in the product stream. The concentration and recovery of methane would, to some extent, be lost either. Therefore, adsorbents prefer to adsorb nitrogen were searched for. It is possible to develop such material taking advantage of steric hindrance. The little difference in molecule diameter between methane (0.382 nm) and nitrogen (0.368 nm) [24] was used to create an adsorbent that allowing for the entrance of N_2 but retarding CH_4. As consequence, methane becomes less adsorbed and it turns out to be the column-top product in a PSA process. Clinoptilolite [25] and its modified species [26] succeeded in switching over the adsorption preference between methane and nitrogen. However, the separation coefficient is still not large enough to have a low cost separation process. It may be more practical to apply an adsorbent that prefers the adsorption of methane but with a high separation coefficient while modification was made on the PSA process having methane as the column-top product.

Another methane separation problem is frequently met in the practice of natural gas exploration. It requests the removal of CO_2. The key components are CH_4/CO_2 then. Many adsorbents have been tested for this separation, but no satisfactory results were reported [27]. Emergence of novel materials promoted the research on CO_2 separation. The MOF (metal organic frameworks)-177 [28] and ZIF (Zeolitic Imidazolate Frameworks)-69 [29] showed a large adsorption capacity for CO_2. However, the adsorption selectivity is not decided by the adsorption capacity of CO_2, but by the difference in adsorption capacity for CO_2 and for other key component gas. In addition, the high manufacture cost of these novel materials must not be acceptable for industrial applications. The role the adsorption mechanism played is especially important for the separation of CH_4/CO_2. The critical temperature of methane is – 82.6 °C, but that of CO_2 is 31.05 °C; therefore, methane is a supercritical gas, but CO_2 is not at usual ambient temperatures. They follow different adsorption mechanisms. While CH_4 is adsorbed on surface of adsorbents merely monolayerly, CO_2 fills in the volume of micropores or condenses in mesopores. As a result, the adsorption difference between CH_4 and CO_2 would not be prominent on microporous adsorbent due to the small pore dimension, however, it becomes prominent on mesoporous adsorbents due to the condensability of CO_2 in the space volume of pores. This argument is proven by the experimental data shown in Table 2. Activated carbon AC-CS was made from coconut shells and possessed plenty of micropores and less mesopores. Activated carbon AC-BB was made from bamboo chips and possessed plenty of mesopores and less micropores. AC-BB has larger adsorption preference for CO_2, though AC-CS has much larger specific surface area and larger adsorbed amounts. Therefore, research efforts should be paid to mesoporous adsorbents for the separation of methane and carbon dioxide.

Methane Storage

Studies on methane storage were initiated by NGV (natural gas vehicle) aiming to decrease the storage cost of CNG (Compressed natural gas). After trying on different materials, the so-called superactivated carbon with as high a specific surface area as over 3000 m^2/g was identified the best adsorbent for methane storage. However, the maximum storage capacity of the ANG (adsorbed natural gas) was more or less 150 V/V compared to 200 V/V of CNG. The sustainable power provided by a tolerable size of fuel tank is still far from that required in driving practice, which is the first drawback of ANG. The second drawback is the large thermal effect on fast charging and discharging, which led to the decrease of storage capacity for about 20 % and difficulty in fast releasing fuel due to the low temperature of adsorbent caused by the fast release. The third drawback is the large residual quantity of fuel that cannot release when the tank pressure dropped to atmospheric. The forth drawback is the difficulty accounted in increasing the package density of superactivated carbon. To reach the storage capacity of about 150 V/V, the superactivated carbon must reach a packing density of about 0.6-0.7 $g.cm^{-3}$, which is difficult especially for large-scale processing. Finally, methane is not same as natural gas, though methane was used to represent natural gas in research. The C_2^+ components of natural gas cannot be desorbed on releasing the storage pressure to atmospheric [30]; therefore, all components other than methane must be removed before storage in order to maintain the storage capacity, which is unreal in practice. Although ANG has been studied for about 30 years, it has never been commercially applied.

Recently, new hope was brought to ANG by novel material. It was reported that a MOF material named PCN-14 possessed extremely large methane uptake capacity, as large as 230 V/V [31]. In fact, the large capacity was a result of misusage of "packing density". MOF is a highly microporous material; therefore, the bulk density is very small. Authors of the paper took the *crystallographic density* for the packing density of the material packed in a fuel tank. The storage capacity of PCN-14 expressed on the basis of unit mass is quite normal as shown in Figure 6, where the adsorbed amounts of methane at 3.5 MPa and 298 K on 16 different kind of adsorbents are correlated linearly with the specific surface area of materials [32]. The materials included are listed in Table 3. The monolayer adsorption mechanism of methane is once more proven in the Figure. According to the mechanism, one must create a material with as large surface area as possible to enhance the storage. However, the specific surface area is not boundless. The theoretically maximal surface area of carbonaceous material applicable for methane storage is only 2620 $m^2.g^{-1}$ [33]. It is, therefore, concluded that any novel material cannot assume the duty of methane or natural gas storage basing on the mechanism of adsorption. Other storage mechanism may bring real new hope for the storage job.

Table 3. Materials presented in Figure 6.

Point	Material	A, $m^2.g^{-1}$	n, $mmol.g^{-1}$
1	MWCNT	137	0.50
2	Zeolite 13X	168	0.77
3	Silica gel	377	0.88
4	MCM41	657	2.10
5	H103[a]	789	3.30
6	ACF1[b]	800	4.92
7	CMK-3[c]	1098	4.98
8	ACF2	1173	5.10
9	ML-L[d]	1215	6.57
10	AC-BY0	1718	8.80
11	PCN-14[e]	1753	9.23
12	AC-LM1[f]	1874	8.60
13	AC-BY1	2063	8.76
14	AC-Paper	2286	8.65
15	AC-LM3	2548	8.90
16	AC-BY2	2800	10.60

Note: a – Adsorption resin made in the Resin Factory of Nankai University, Tianjin, China; b – ACF = activated carbon fiber; c – A carbon mesoporous molecular sieve, refer to (Liu et al, 2006) for details of structure; d – A hypercrosslinked polymer network, refer to (Wood et al, 2008) for details of structure; e – A MOF material, refer to (Ma et al, 2008) for details of structure; f – AC = activated carbon, either purchased or made in the author's lab.

Providing the condition of forming hydrates in the pore space of materials, the author and coworkers proposed a wet storage method for methane and natural gas [34]. A definite amount of water was pre-adsorbed in adsorbents, and then methane or natural gas was charged in. The wet storage method shows advantages in comparison with ANG. First, the storage capacity can be as large as CNG while the storage pressure is merely one third of the latter; second, the thermal effect on charging and discharging doesn't affect the storage capacity or fast discharge fuel; third, the residual capacity at ambient pressure is almost zero; fourth, components of natural gas below C_4 can form hydrates; therefore, only C_5^+ should be removed before charge, which is easily done; fifth, wet carbon powder is plastic; therefore, its packing density can be easily tuned to the optimal value [35]. After a systematic study, it was realized that mesoporous carbon with pore dimension 2-5 nm is preferred for the wet storage material, and as high as 41.2 wt % capacity was reached [36]. So, different mechanism calls for different materials of storage. Advantage of the wet storage method over the HNG (hydrate of natural gas) lies on the fact that both conversion and dynamics of forming hydrates in water media is not satisfactory, whereas 100 % conversion and fast formation were observed in the pore spaces [37].

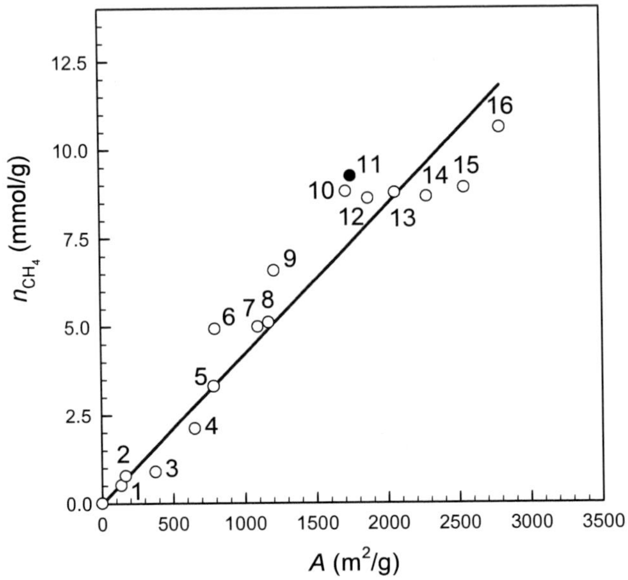

Figure 6. Methane adsorption on different materials at 3.5 MPa and 298 K.

CONCLUSION

To develop a method or a material targeting at methane separation or storage basing on the principle of adsorption, the mechanism of adsorption must be considered. Methane adsorbs on the surface of any sort of materials merely monolyerly unless the temperature of adsorption is below -82.6 °C. According to the adsorption mechanism, any materials cannot assume the duty of storing methane or natural gas based on the principle of adsorption. Forming hydrates in the pore spaces of carbon materials provides another storage mechanism that seems more practical. However, different mechanism calls for different materials for storage. Understanding adsorption mechanism may also drop a hint on enhancing the separation of methane from air and carbon dioxide.

ACKNOWLEDGMENTS

Financial support of the National Natural Science Foundation of China (grant number 90510013 and 20336020) is greatly acknowledged.

REFERENCES

[1] Pelley, J. ((2008)). Gas hydrates on the front burner. *Environ. Sci. Technol, ASAP Web* Release Date: 27-Aug-2008 (News) DOI: 10.1021/es802250e.

[2] Gregg, S. J. & Sing, K. S. W. (1982.). *Adsorption Surface Area and Porosity*. 2nd ed. London: Academic Press.

[3] Malbrunot, P., Vidal, D., Vermesse, J., Chahine, R. & Bose, TK. (1992). Adsorption measurements of argon, neon, krypton, nitrogen, and methane on activated carbon up to 650 MPa. *Langmuir, 8,* 577-580.

[4] Aranovich, G. L. & Donohue, M. D. (1996). Adsorption of supercritical fluids, *J. Colloid. & Interf. Sci., 180,* 537-541.

[5] Bénard, P. & Chahine, R. (1997). Modeling of High Pressure Adsorption Isotherms above the Critical Temperature on Microporous Adsorbents: *Application to Methane. Langmuir, 13,* 808-813.

[6] Ustinov, E. A. Do, D. D., Herbst, A., Staudt, R. & Harting, P. ((2002). Modeling of gas adsorption equilibrium over a wide range of pressure: A thermodynamic approach based on equation of state. *J Colloid & Interf Sci., 250,* 49-62.

[7] Neimark, A. V. Ravikovitch, P. I. (1998). *Calibration of adsorption theories.* In Meunier F (ed). FOA6 Proceedings. Paris: Elsevier, 159-164.

[8] Everett, D. H. (1971). *Manual of symbols and terminology for physicochemical quantities and units.* Appendix II. Part I, London: Butterworth.

[9] Zhou, L. (2002). Chapter 4: Adsorption Isotherms for the Supercritical Region. In Toth J, editor. Adsorption: *Theory, Modeling & Analysis.* New York: Marcel Dekker, 211-250.

[10] Zhou, L., Bai, S. P., Zhou, Y. P. & Yang, B. (2002). Adsorption of Nitrogen on Silica Gel over a Large Range of Temperatures. *Adsorption, 8,* 79-87.

[11] Murata, K. & Kaneko, K. (2000). Nano-range interfacial layer upon high-pressure adsorption of supercritical gases. *Chem Phys Letters, 321,* 342-348.

[12] Humayun, R. & Tomasko, D. L. (2000). High-Resolution Adsorption Isotherms of Supercritical Carbon Dioxide on Activated Carbon. *AIChE J, 46,* 2065-2075.

[13] Zhou, L. & Zhou, Y. P. (1998). Linearization of Adsorption Isotherms for High Pressure Applications. *Chem Eng Sci., 53,* 2531-2536.

[14] Zhou, L. Zhang, J. S. & Zhou, Y. P. (2001). A Simple Isotherm Equation for Modeling the Adsorption Equilibria on Porous Solids over Wide Range Temperatures. *Langmuir, 17,* 5503-5507.

[15] Ruthven, D. M. (1984). *Principles of Adsorption and Adsorption Processes*. New York: John Wiley & Sons.

[16] Zhou, L., Zhou, Y. P., Li, M., Chen, P. & Wang, Y. (2000). Experimental and Modeling Study of the Adsorption of Supercritical Methane on a High Surface Carbon. *Langmuir, 16,* 5955-5959.

[17] Zhou, L., Zhou, Y. P. & Sun, Y. (2006). Studies on the mechanism and capacity of hydrogen uptake by physisorption-based materials. *Int. J. Hydrogen Energy, 31,* 259-264.

[18] Zhou, L., Bai, S. P., Su, W., Yang, J. & Zhou, YP. (2003). Comparative Study of the Excess versus Absolute Adsorption of CO_2 on Superactivated Carbon for the Near-Critical Region. *Langmuir, 19,* 2683-2690.

[19] Schneider, M. S., Grunwaldt, J. D. & Baiker, A. (2004). Near-Critical CO_2 in Mesoporous Silica Studied by In Situ FTIR Spectroscopy. *Langmuir, 20,* 2890-2899.

[20] Ruthven, D. M. (2000). Past Progress and Future Challenges in Adsorption Research. *Ind. Eng. Chem. Res., 39,* 2127-2131.

[21] Gu, M., Xian, X. F., Zhang, D. J. & Chen, C. G. (2002). *Separation of CH_4/N_2 mixtures by pressure swing adsorption.* J Coal. (in Chinese*), 27,* 197-200.

[22] Yue, Y. H., Tang, Y. & Gao, Z. (1995). Modification on zeolite openings and adsorptive separation of gases. *Acta Physico-Chimica Sinica, 1,* 912-915.

[23] Zhou, L., Guo, W. C. & Zhou, Y. P. (2002). A Feasibility Study of Separating CH_4/N_2 by Adsorption. *Chn J. Chem. Eng., 10,* 558-561.

[24] Hirschfelder, J. O. Curtis, CF; Bird, RB. (1954). *Molecular Theory of Gases and Liquids.* New York: Wiley.

[25] Jayaraman, A., Hernandez-Maldonado, A. J., Yang, R.T., Chinn, D., Munson, C. L., Donald, D. H. & Mohr, H. (2004). Clinoptilolites for nitrogen/methane separation. *Chem Eng Sci., 59,* 2407-2417.

[26] Jayaraman, A. & Yang, R. T. (2005). Tailored Clinoptilolites for Nitrogen/Methane Separation. *Ind.* Eng. *Chem. Res., 44,* 5184-5192.

[27] Abanades, J. C., Rubin, E. S. & Anthony, E. J. (2004). Sorbent Cost and Performance in CO_2 Capture Systems. *Ind. Eng. Chem. Res., 43,* 3462-3466.

[28] Millward, A. R. & Yaghi, O. M. (2005). Metal-Organic Frameworks with Exceptionally High Capacity for Storage of Carbon Dioxide at Room Temperature. *J. Am. Chem. Soc., 127,* 17998-17999.

[29] Banerjee, R., Phan, A., Wang, B., Knobler, C., Furukawa, H., O'Keeffe, M. & Yaghi, O. M. (2008). High-Throughput Synthesis of Zeolitic Imidazolate Frameworks and Application to CO_2 Capture. *Science, 319,* 939-943.

[30] Fu, G. Q. & Zhou, L. (2000). Experimental study on adsorptive storage of natural gas. I. Effect of minor content of ethane on storage capacity. Chem Technol Natural Gas (in Chinese). 2000, 25, 12-14; II. Effect of minor content of propane and butane on storage capacity. *Chem Technol Natural Gas, (in Chinese), 25,* 22-24.

[31] Ma, S. Q., Sun, D. F., Simmons, J. M., Collier, C. D., Yuan, D. Q. & Zhou, H. C. (2008). Metal-Organic Framework from an Anthracene Derivative Containing Nanoscopic Cages Exhibiting High Methane Uptake. *J. Am. Chem. Soc., 130,* 1012-1016.

[32] Sun, Y., Liu, C. M., Su, W., Zhou, Y. P. & Zhou, L. Principles of Methane Adsorption and Natural Gas Storage. *Adsorption*, in press.

[33] Matranga, K. R., Myers, A.L. & Glandt, E. D. (1992). Storage of natural Gas by Adsorption on Activated Carbon. *Chem Eng Sci., 47,* 1569-1579.

[34] Zhou, L., Sun, Y. & Zhou, Y. P. (2002,). Enhancement of the Methane Storage on Activated Carbon by Pre-adsorbed Water. *AIChE J., 48,* 2412-2416.

[35] Zhou, Y. P., Wang, Y. X., Chen, H. H. & Zhou, L. (2005). Methane storage in wet activated carbon: Studies on the charging/discharging process. *Carbon, 43,* 2007-2012.

[36] Liu, X. W., Zhou, L., Li, J. W., Sun, Y., Su, W. & Zhou, YP. (2006). Methane Sorption on Ordered Mesoporous Carbon in the Presence of Water. *Carbon, 44,* 1386-1392.

[37] Zhou, Y. P., Dai, M. & Zhou, L. (2004). Storage of Methane on Wet Activated Carbon: Influence of Pore Size Distribution. *Carbon, 42,* 1855-1858.

In: Energy Costs, International Developments…
Ed: Leszek Kowalczyk and Jakub Piotrowski
ISBN: 978-1-60741-700-2

Chapter 8

PERSPECTIVE OF ENERGY AND GHG EMISSION IN CHINA IN 2050

Masanori Tashimo*
Energy Think Tank Co. Ltd. Japan, 4-7-13 Setagaya-ku, Tokyo, Japan

ABSTRACT

During the past several years, world climate conditions have been abnormal. The G8 summit was held at Lake Touya in June 2008. At this meeting the G8 leaders agreed to make efforts to reduce CO_2 emissions to half of the present amount by 2050. The IPCC's studies of world energy consumption and emissions of CO_2 in 2050 are based on SRES-B CO_2, based on moderate world economic growth. CO_2 emissions are estimated to be 12.3Gt-C in 2050.

However, Asia's economic growth rate was higher than expected in recent years. At present, energy consumption in China in 2050 is expected to be 6Gtoe based on the past trend of Japan's and Korea's economic growth.

In the case of a 450ppm limitation, the allowable world CO_2 emission is about 4.8GtC by 2050. China's CO_2 emission is estimated to be 4.1Gt-C based on 6Gtoe energy consumption with the same IPCC t-C/toe ratio. A reduction of one-fifth of CO_2 emissions will be required in China by 2050.

There are two ways to reduce CO_2 emissions: one is reduction of energy consumption and another is the use of less CO_2 emission energy resources.

Concerning energy consumption, the ratio of energy consumption in Japan and China in the industrial field is about 1.5. Furthermore, it is required that energy efficiency will be improved two times the present rate of efficiency by 2050 in all fields of energy supply and consumption. The ratio of CO_2 emission per energy, t-C/toe, should be reduced to about half by substituting renewable energy (such as hydro, wind, photovoltaic, biomass), nuclear energy and CCS.

There are difficult barriers to achieving this target. The key components of this achievement are technologies in efficiency improvement and less CO_2 emissions in the energy supply at reasonable cost to assure the economic development. Also, the political stability and regimes are essential matters domestically and internationally. The climate

* Corresponding author: E-mail ett@amber.plala.or.japan

crisis is a critical issue for mankind. It is expected that China will play a significant leadership role in the world.

1. Preface

During the past several years, the economic growth in the world took off to a very high level, especially in Asia (Figure1-1). There has been a possibility of a great inflation since the sixteenth century. Worldwide sliding stock prices have occurred because of the United States' subprime loan problem resulting in the bankruptcy of the Lehman Brothers bond.

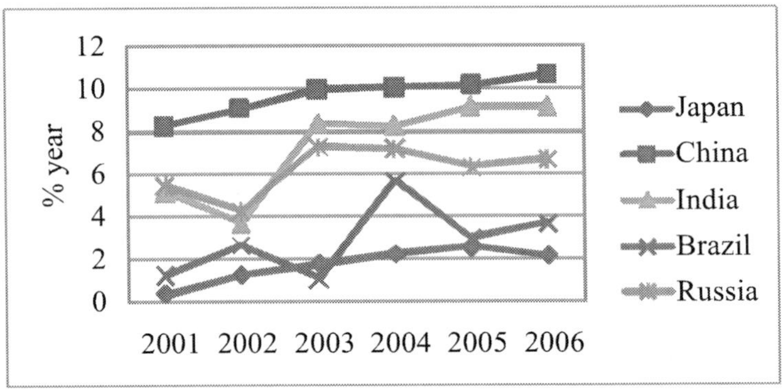

Figure 1-1. GDP Growth Rate

Moreover, oil prices have trended downward to $50/bal from $140/bal at last September. These two issues have contributed to the crash of the financial industry resulting in an increase of the world GDP of more than 4%. World economic growth is expected to stabilize, and many countries will strengthen climate issue activities and sound economic development. China's efforts on the greenhouse gas (GHG) issue is a key international issue.

2. Kyoto Protocol and G8 Hokkaido Toyako Summit Leaders Declaration

(1) Kyoto Protocol

The Kyoto Protocol was signed by many countries in 1997. There was a consensus that climate change is a dire issue for the survival of the human race and civilization. There was also a consensus that developed countries were responsible for the increase in GHG in the air up to the present.

At the time the IPCC Third Assessment Report was published, there was a consensus that we should keep the CO_2 concentration under 550ppm in the future. Corresponding to this target, a Clean Development Mechanism (CDM) method was selected as a cheaper method to reduce world CO_2 emissions. At present, the authorized CDM projects number over 1000.

(2) G8 Hokkaido Toyako Summit Leaders Declaration

In the G8 Summit Leaders Declaration of 2008, it was agreed that we should reduce CO_2 emissions to about a half of present emissions. This corresponds to an increase in temperature of around two degrees from present levels and the CO_2 concentration should be stabilized under 450ppm in the future. This conclusion was strongly affected by the IPCC's fourth report in 2007. In this report, it is pointed out that the early effort to reduce CO_2 emissions will be very effective in protecting the investment and the cost of recovery from climate change.

There are large differences between a 550ppm limitation and 450ppm limitation. The CO_2 concentration before the industrial revolution in the nineteenth century was stable at 280ppm level. The CO_2 concentration at present is about 380ppm. The CO_2 concentration has already increased by 100ppm. If the upper limitation is assumed to be 550ppm, it is permitted to emit CO_2 at about a 170ppm increment. However, if it is assumed to be 450ppm, it is permitted to emit CO_2 by only an 80ppm increment. This means that the allowable CO_2 emission is based on levels already emitted to the atmosphere since the nineteenth century.

There have been several studies on predicted future CO_2 emissions. The WRE report has been used by many researchers. In this report, allowable CO_2 emissions by 2050 are only 4.8Gt-C and must be decreased more after 2050 (Figure 2-1).

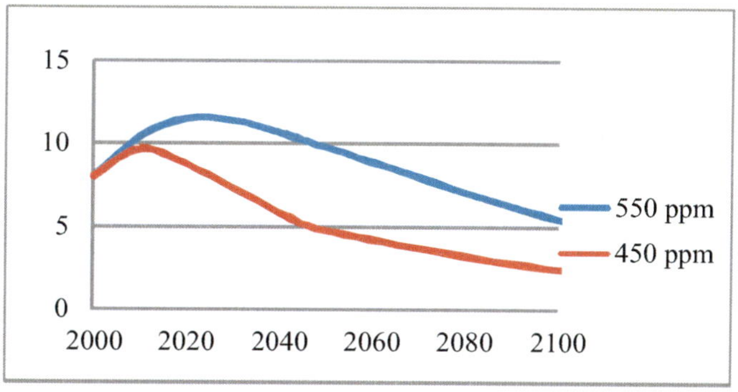

Figure 2-1.WRE CO2 Limits

(3) The Impact of Cutting Present CO_2 Emission Levels in Half

In the case of 550ppm, there are time margins for achieving the target of relatively high allowable CO_2 emissions over next several decades. The CO2 emissions in OECD countries is restricted and the CDM was selected to mitigate the increased rate of CO_2 emissions in developing countries. However, in the case of 450ppm, there is little time to reduce world CO_2 emissions not only in OECD countries but also in developing countries. In this sense, a CO_2 emission trade method will be less effective in reducing CO_2 emissions.

(4) CO_2 Emissions

The energy demand essentially increases as GDP increases in modern society. CO_2 emission increases as energy consumption increases. However, hereafter it is necessary to reduce CO_2 emissions while energy consumption increases.

The IPCC estimated CO_2 emissions in 2050, shown in Figure 2-2. This estimation is based on SRES-B CO_2. As shown in Figure 2-2, CO_2 emissions in 2050 will be 12.3Gt-C, which is 2.5 times the allowable CO_2 emissions under the 450ppm limitation. It is required to reduce CO_2 emissions not only in advanced countries up to -70% of expected emissions by 2050, but also in developing countries up to -60% of expected CO_2 emissions by 2050 (Figure 2-3, Table 2-1).

According to IPCC estimation, CO_2 emissions in China is 19% of total world emissions. The allowable CO_2 emissions in China will be 0.91Gt-C in 2050 based on this share.

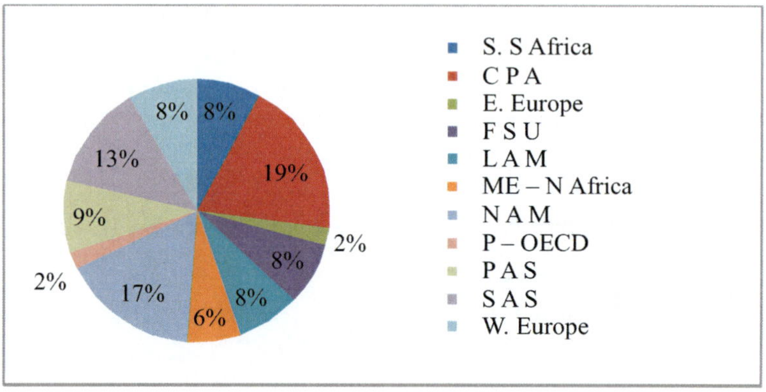

Figure 2-2. CO2 emission at 2050 Total 12.3 Gt-C

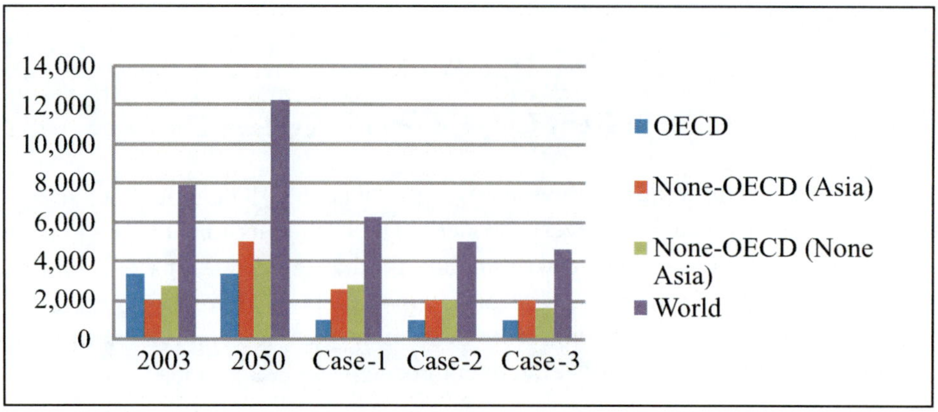

Figure 2-3. CO2 Emission and Reduction
Based on SRES-B by IPCC (Mt-C)

Table 2-1. CO2 Reduction Rate

	Case-1	Case-2	Case-3
OECD	0.3	0.3	0.3
None-OECD (Asia)	0.5	0.4	0.4
None-OECD (Not Asia)	0.7	0.5	0.4

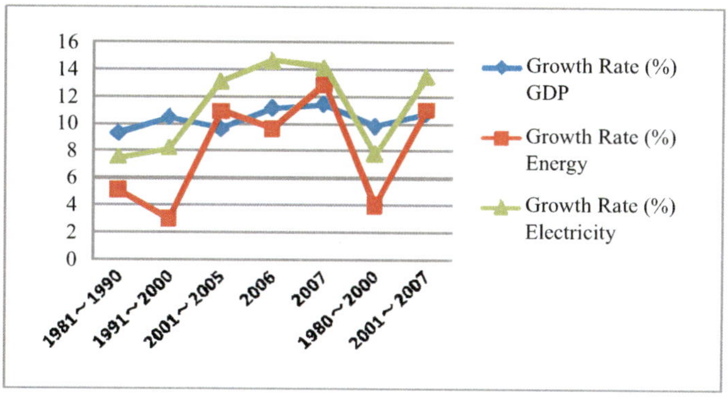

Figure 3-1. Increase Rates in Economy, PE and EL

3. Perspective of Energy in China

(1) Trend Survey of China in the Past 20 Years

Economic growth in China was high during the last 20 years. The growth rate of GDP, Primary Energy (PE) and Electricity (EL) is shown in Figure 3-1. These were especially high during 2000 to 2007. China was already in a takeoff situation. This extremely high increase in the economy induced the international inflation of natural resources and resulted in worldwide high economic activity during the last two to three years.

(2) Past Trend of Japan and East Asian Countries

Takeoff in Japan started in 1960 and high economic growth continued until the time of the oil crisis in 1973. During this period, average GDP growth rate was about 8% per year. Following Japan, takeoff started in Korea and Taiwan in the 1980s. The annual increase rates of GDP, PE and EL in these three countries are shown in Figure 3-2, 3-3 and 3-4, respectively.

The annual PE increment in Japan was down to about 0% during 1973 to 1985 from 8% during 1960 to 1973. However the annual GDP increment in Japan was decreased to 3–4% from 8% during the same time by efforts in energy savings and the industrial structural change, for example, from heavy industries to light industries.

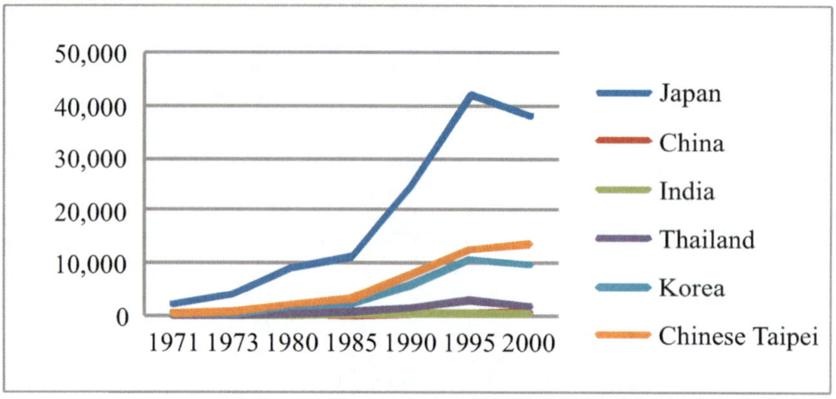

Figure 3-2. GDP/capita($) in Asia (1971-2000)

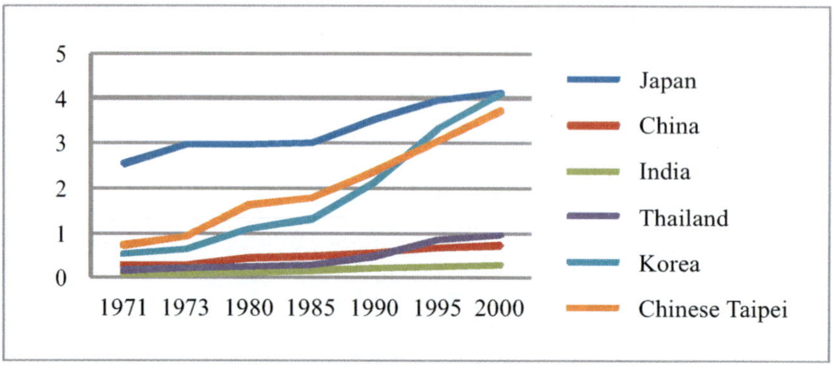

Figure 3-3. Energy/capita(toe) in Asia (1971-200)

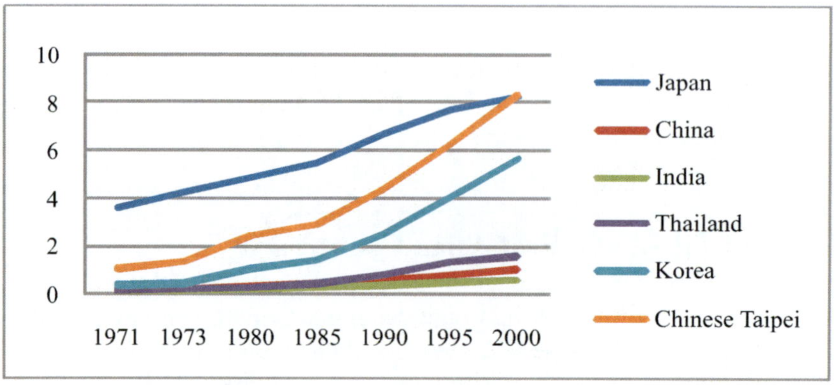

Figure 3-4. Electricty/capita(1000KWh) in Asia (1971-2000)

In East Asia, Korea and Chinese Taipei were at the takeoff stage in economic development during the 1970s. In these countries, the increase in GDP, PE and EL rates were not as greatly affected by oil crises as Japan. In the 1980s and 1990s, the economic growth in these countries continued. However the rate increase was gradually decreasing. Concerning EL in Japan, the increase rate was decreased after 1973, but it did not decrease as much as

GDP and PE because of the convenience of electricity especially in residential and commercial use.

At present, EL per capita is almost the same as that of the OECD EU. EL per capita in Korea and Chinese Taipei has risen constantly since the 1980s and has reached close to OECD levels. These trends in the three countries show a typical pattern of economic development. The trend in Japan shows a pattern ranging from a period of high economic development to a stable economic period ending in a stable economic development period. The trend in Korea and Chinese Taipei shows a pattern from a pre-high economic growth period to a high economic growth period and finally to a stable economic development period.

In Figure 3-5 and Figure 3-6, the relationship between GDP, PE and EL are shown. These figures show the same trend: PE and EL increase as GDP increases. At a low level of GDP, the PE and EL increase rate is very high. But these increase rates slow down over 10,000$/c. This trend is expressed as a formula of y=A ln(x) +B in a simplified equation.

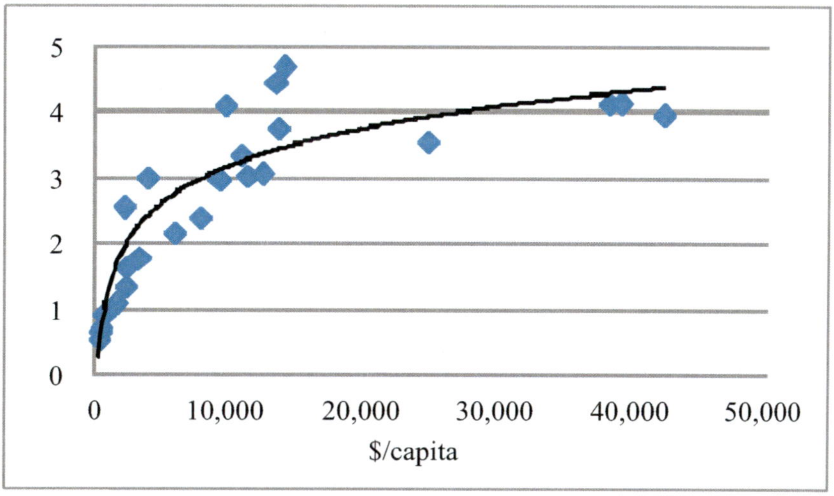

Figure 3-5. P. Energy(toe)/c vs GDP/c in Asia

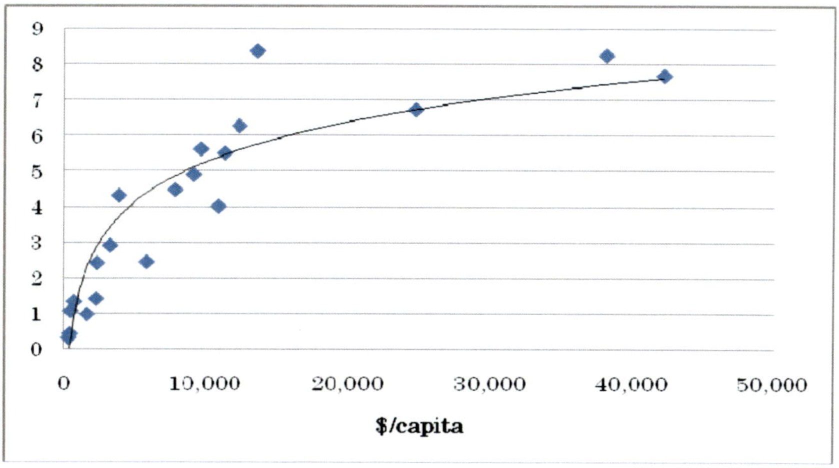

Figure 3-6. Electricty(1000KWh)/cvs GDP/c in Asia (Revised)

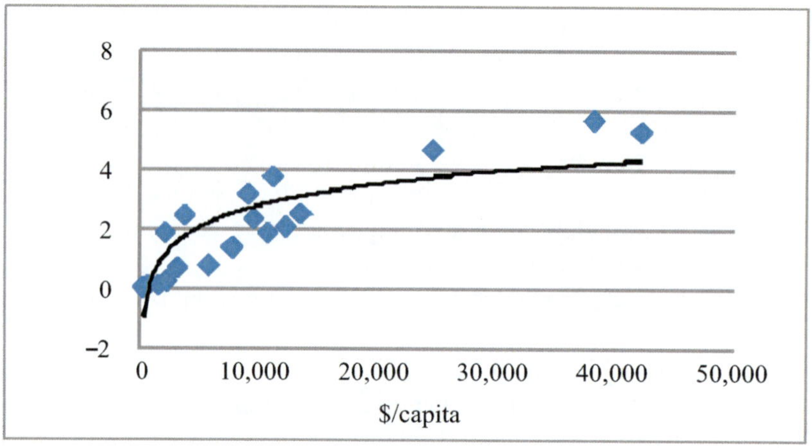

Figure 3-7. Car penetration(Cars/10 capita) vs GDP/c in Asia

To estimate energy demand in China, the energy demand for transportation is very important, especially considering the energy for automobile as a dominate factor in the future. To estimate the energy for transportation, the trend of market penetration of cars is shown in Figure 3-7.

(3) Perspective of Energy Demand in China

There are several long-term energy perspectives. The SRES series was used by many researchers. This was seen as an underestimated perspective in Asia, especially the Chinese energy demand to 2050. This perspective of world energy demand is about 20 Gtoe and that of China is 3.55 Gtoe. There was a perspective of energy in China to 2050 which was based on *Dreaming with BRICs: The Path to 2050* by Goldman Sachs (ref.). In this report, a new Chinese perspective of energy is discussed. The perspective of GDP is based on Goldman Sachs' *Dreaming with BRICs* (see Figure 3-8). The perspective of PE and EL are based on Asian trends, as discussed in the previous section. The results of PE and EL are shown in Figure 3-9 and Table 3-2.

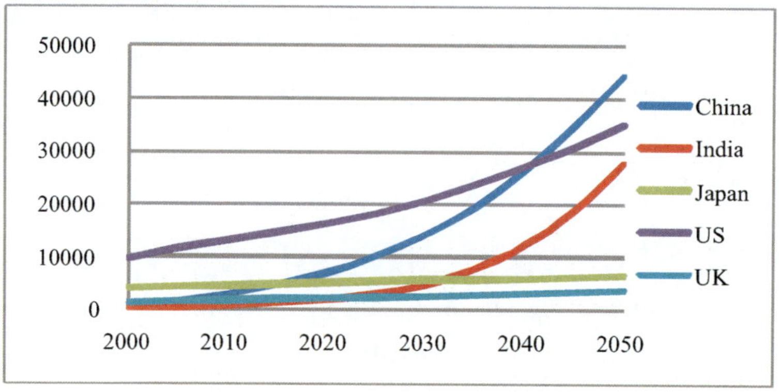

Figure 3-8. GDP perspective(B$) by Goldman Sachs.

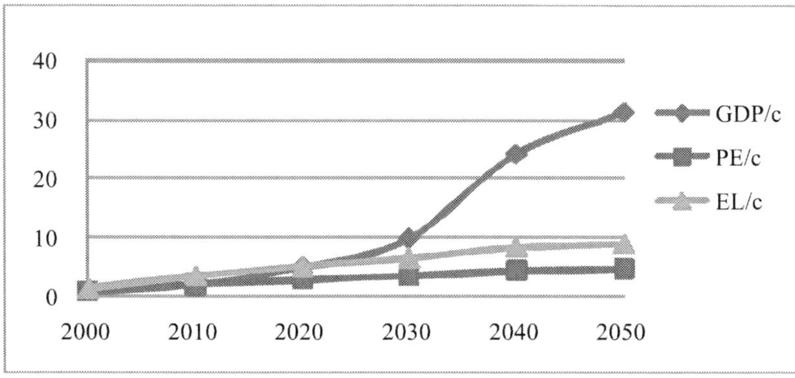

Figure 3-9. Perspective GDP/c(1000$), PE/c(toe), EL/c(MWh) in China 2000-2050

Table 3-1. GDP/c,PE/c, EL/c, Car/10c Perspective in China (Revised)

	2000	2010	2020	2030	2040	2050
GDP/c	0. 845	2. 233	4. 965	9. 809	24. 192	31. 357
PE/c	1. 01	2. 01	2. 84	3. 54	4. 47	4. 74
EL/c	1. 24	3. 48	5. 14	6. 55	8. 41	8. 95
Car/10c	0. 16	1. 2	2	2. 8	3. 7	4

The PE and EL perspective are about 3toe/c and 5300KWh/c and the total demands of these are 4.5Gtoe and 8PWh, respectively by 2030. In 2050, these are estimated to be about 6Gtoe and 10.8PWh, respectively. There is another way to estimate the energy demand—by the elastic factor method—resulting in 5.6–6.6Gtoe for total energy demand. Whichever method is selected, the total energy demand in China is about 1.5 times larger than SRES-B.

Concerning the energy demand for transportation, the penetration of the car is expected to be 2.8 cars/10 persons in 2030 and 4 cars/10 persons in 2050. These results show that the total number of cars in China will be 400 million in 2030 and 600 million in 2050.

4. Discussion: What Shall Be Done in China?

(1) General Discussion

As discussed above, there is a risk that CO_2 emissions in China will be as great as the 2050 total world allowable CO_2 emission. Therefore, it is expected that China shall play the most important role in environmental issues in the world. The economic growth of a big country with a vast population will affect the world economy more than other countries, such as Japan.

In addition to China, India and ASEAN countries are expected to take off in the near future. There is a possibility that a great age of inflation will occur as was experienced in the sixteenth century. In addition to this inflation issue, we have severe restrictions of environment. This is first restriction of human beings in modern civilization.

There is a possibility that China's CO_2 emissions will be as great as the total allowable world CO_2 emissions in 2050 according to the past relation between GDP and Primary Energy, as discussed in section 3. This section discusses what can be done to avoid this situation in future. The main discussion is focused on three technical issues: electricity, transportation and nuclear energy. The energy consumption in electricity and transportation is more than a half of the total energy consumption in advanced countries at present. Then discussions follow regarding how to reduce CO_2 emission in electricity and transportation, and the role of nuclear energy.

Concerning reduction of CO_2 emissions, the technical issues are a key factor. However, the spread of new technology is also an important factor, especially in big countries such as China. To promote the spread, the social system of a country and a region are also vital issues; for example, the tax system in software and the public transportation system in hardware.

(2) Discussion of Electricity

Concerning the reduction of CO_2 emissions, there are two approaches in electricity generation: quality and quantity. In this section, case studies are done within a possible range.

Case Selection

The present resources of Chinese power plants are shown in Table 4-1 (P2.5). In 2005, electricity generation by coal fire was about 80%, by hydro about 15% and by nuclear plants about 5%.

The total electricity generation was 2.5PWh in 2005.The total electricity generation is estimated 10.8PWh in 2050. In a qualitative discussion, two cases are selected: 10.8PWh cases, in which P10.8 is the most conservative case and F10.8 is the desired future case using the hydro (10%), renewable energy resources (20%) and nuclear plants (20%). In a quantitative discussion, a 10.8PWh case and an 8.1PWh case (P8.1) are selected.

Table 4-1. Case of Electric Generation (PWh) (Revised)

	2005	2050	2050	2050	2050
	P2.5	P10.8	F10.8	F8.1	FA8.1
Hydro	0.397	1.74	1.08	0.81	0.81
Fossil	2.02	8.84	4.32	3.24	2.43
Nuclear	0.053	0.09	3.23	2.43	3.23
Renewable E.	0	0	2.16	1.62	1.62
Sum	2.469	10.8	10.8	8.1	8.1

The latter one would yield a 25% reduction in electricity demand by improvements in efficiency and saving. In addition to these cases, the advanced case (FA8.1) is studied to reduce CO_2 emissions by nuclear (+10%) replacing the coal fired thermal plants (-10%). The details are shown in Table 4-1 and Figure 4-1.

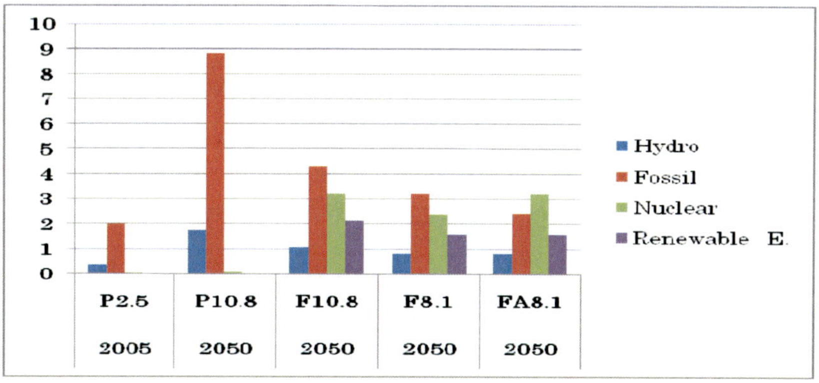

Figure 4-1. Perceptive of Electricty in China in 2050 Case Condition (PWh) (Revised)

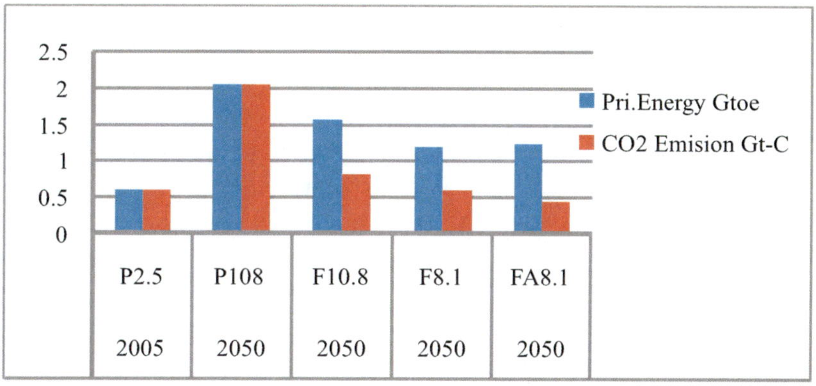

Figure 4-2. PE and Emission in China in 2050

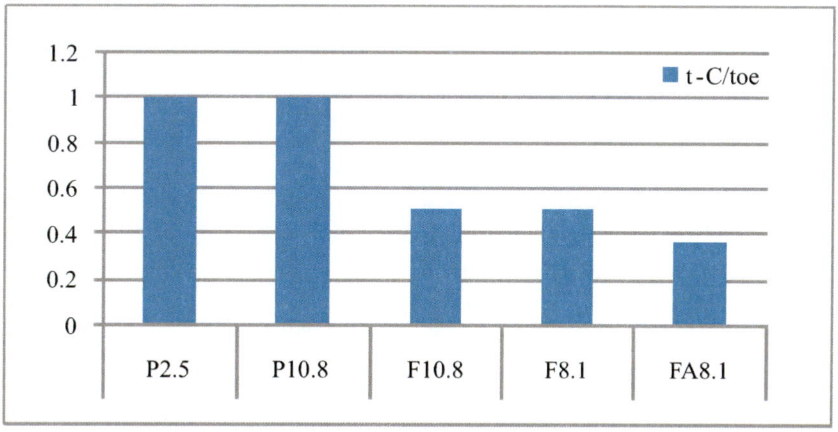

Figure 4-3.Emission Rate of EL in China in 2050

In China, hydro power is expected to be 2~3PWh in the southeast area. If hydro power is constructed as expected, it will reduce the burden to renewable energy duty.

Major Conclusion

a. In case of P10.8, it is certain that the CO_2 emission by coal fire thermal plants is estimated to be 2.2Gt-C, which is one-half of the allowable CO_2 emission of 4.8 Gt-C in the world by 2050.
b. In case of F10.8 which is a candidate for future formation of resources of electricity generation, the CO_2 emission by electricity generation will reduce to 0.8Gt-C. In this case, the efficiency of coal fire power plants is assumed to be 50% by the IGCC.
c. The CO_2 emission of F8.1, which is the case of 25% reduction of electricity demand, is estimated to be 0.6Gt-C. The CO_2 emission of FA8.1, which implies a 10% nuclear increase and 10% coal fire plant decrease, is estimated to be 0.46Gt-C, which is one-quarter of P10.8.
The temporary target amount of CO_2 emissions in China by 2050 is assumed to be 0.91Gt-C in this study. This amount is based on the same share (19%) of IPCC CO_2 perspective under the scenario B CO_2 in 2050. The amount of CO_2 emissions of F8.1 is 0.46Gt-C, about 50% of the tentative allowable CO_2 emissions in China.
d. In a qualitative aspect, regarding the CO_2 emission rate per unit of energy—say t-C/toe—less is better. At present, it is 1t-C/toe in the case of P2.5. In the case of FA8.1, it is reduced to 0.4t-C/toe by renewable and nuclear energy. This means that half of the reduction of CO_2 emission is achieved by qualitative improvement. In addition to this, there is a way of CCS to improve energy quality, even though it may be expensive.

(3) Discussion of Energy for Transportation

Energy for transportation was 0.114Gtoe in 2005. This is mostly used by railway and truck for transporting material. This is a share of 6.6% and is smaller than the world average share of 19%. In the future, the automobile is expected to increase up to 4 billion by 2050 with the improvement of standards of living. The increment will result in an increase in energy for transportation of up to 0.76 Gtoe, and 0.61 Gt-C for CO_2 emissions. This estimation is based on 1.9toe/car which is the average energy consumption in the world at present.

By this estimation regarding cars, it is expected that the energy for transportation of Primary Energy base will be 0.95 Gtoe (0.76/0.8) including the refining and distribution (from factory to stand) of energy. This means that the CO_2 emissions for transportation will overtake the emissions of the advanced electric generation by 2050.

At present, the ICE (Internal Combustion Engine) is usually used for automobiles. The ICE efficiency is rather poor—as much as 15%. The efficiency of the hybrid car (HC) is about 30%. This type of car is already commercialized in the market, and the electric car (EV) is expected to be more efficient by up to 30% and is now commercialized by many companies. But in the case of EV, CO_2 emission strongly depends on the resources of power plants. In this case, the estimation of CO_2 emissions is based on average of coal fire and natural gas fire power plants.

In the near future, fuel cell vehicles will be commercialized to reduce CO_2 emissions by using hydrogen fuel. The typical energy and emission performances are shown in Table 4-2.

The automobile in China will penetrate very rapidly in the next few decades. The political realm plays a very important role in promoting the popularity of high efficiency automobile use, such as a high penalty for high-emissions cars and a bonus for low-emission cars. By these measurements, it is expected that the CO_2 emission of transportation will be reduced to 0.38Gt-C from 0.76Gt-C and the energy demand of transportation will be reduced to 0.475Gtoe from 0.95Gtoe. In Japan, the average energy consumption is about 1~1.2 toe/car/year, which is half of the world average due to smaller cars and the popularity of the public transportation system. These two factors will be important in saving energy and reducing CO_2 emissions. It is hoped that the CO_2 emissions of transportation will be reduced to 0.2Gt-C and the energy of transportation will be reduced to 0.25Gtoe by 2050.

Table 4.2. Performance of automobile

	Fuel	Consumption (10Km)	CO2Emission Kg/10Km)
ICE	Gasoline	1L	1
Hybrid Car	Gasoline	0.5L	0.5
Electric Vehicle	Electricty	2KWh	0.39
Electric Vehicle	Electricty (nuclear)	2KWh	0
Fuel Cell Vehicle	Hydrogen	1Nm3	0.33

(4) Discussion of Nuclear Energy

Even though it has been pointed out that nuclear energy has issues of proliferation of nuclear weapons and waste management, nuclear energy is expected to play a very important role not only in China but also in Asia to resolve the economic growth and climate issues. It is certain that the energy demand in Asia will increase in the next several decades driven by increases in the standard of living.

2.42 PWh in F8.2 and 3.23 PWh in FA8.2 correspond to the capacity of nuclear plants, 310GWe and 460GWe. At present, under the Chinese Government plan, the capacity of nuclear plants will be 60 GWe in 2020 and 160GWe in 2030. In this case study, the electricity generated by nuclear plants is expected to be 3.23PWh in FA8.2GWe nuclear power plants by 2050. With this nuclear energy, the CO_2 emissions could be reduced up to 0.46Gt-C. There is another choice for reducing CO_2 emissions, i.e., CCS (carbon dioxide capture and storage) for the coal-fired plants. However, CCS method requires the energy to collect and liquefaction of CO_2 gas. At least 25% of generated energy is needed to collect and liquefaction of CO_2 gas. This means that a 50% thermal efficiency plant with IGCC can generate a maximum of 37.5% electricity for service. Also, the transportation of energy to the storage region is required. Concerning CCS, the economy will be an issue.

The renewable energy is favorable for CO_2 emissions. The power availability of wind and photovoltaic is 25% and 13%, respectively. Also, the output of power depends on weather conditions, resulting in the power level to drift from time to time. These issues affect the economy. From these points of view, nuclear energy will be a promising candidate for a stable, central and large scale electric generation system.

In the latter half of this century, it must be maintained that CO_2 emissions should be less than half of present emissions to establish a sustainable economic system. From this point of view, the more effective use of U resource and an easy system (less reprocessing requirement) of nuclear utilization, i.e., "Candle" shall be studied.

(5) Total Energy and GHG in 2050

Based on the present conditions, the total PE for EL and transportation is 3.3Gtoe and CO_2 emissions will be 3.05Gt-C by 2050. As discussed above, these are reduced to 1.55Gtoe and 0.66Gt-C by the shift of energy resources and efficiency improvements. In this section, the industrial, residence and business fields (Others in Table 4-3) are not discussed. However, the PE for these fields will be reduced to 1.45Gtoe from 2.7Gtoe of present estimation. As a result, the CO_2 emissions will be reduced to 0.25Gt-C from 1.05Gt-C (Table 4-3).

Table 4-3. Total Energy and GHG at 2050

	Perspective		Advaced Perspective	
	PE.(Gtoe)	CO2(Gt-C)	PE.(Gtoe)	CO2(Gt-C)
IPCC/World	20	12.3		4.8
/China	3.55	2.6		0.91
ETT/China				
Total	6	4.1	3	0.91
EL	2.1	2.1	1.25	0.46
Trans.	1.2	0.95	0.3	0.2
Others	2.7	1.05	<1.45	<0.25

It is concluded that there will be a possibility to clear the CO_2 tentative limitation by means of technical and institutional efforts.

5. CONCLUSION

(1) It has been pointed out that predictions of energy demand in China will be more than SRES-B CO_2 under stable political conditions and sound economic management. This shall be recognized throughout the world in order to have a reasonable and rational solution to the GHG problem. In this chapter, the perspective of energy in China is estimated by the past trend of three East Asian countries. The PE (Primary Energy) and EL (Electricity) are estimated to be about 6Gtoe and 10.8PWh, respectively. Also, CO_2 emissions are estimated to be 4Gt-C in China by 2050. This emission almost corresponds to the upper limit of world emissions in 2050.

(2) It is expected in electric generation that the CO_2 emissions can be reduced to one-third of present estimation by increasing the renewable energy and nuclear energy, replacing the coal- fired plants. In addition to a qualitative approach, a 25% reduction in electricity demand will reduce CO_2 emissions, resulting in one-quarter of present estimates of the total CO_2 emissions from electricity generation.

The energy for the transportation field is expected to increase rapidly in next few decades by the market penetration of the automobile. However, the more efficient vehicles, such as hybrid cars and EV (electric vehicles) can save half of that level of energy. And there are other ways to save energy in transportation, such as more widespread us of smaller cars and a convenient public transportation system.

The two major fields, i.e., electricity generation and transportation, occupy over half of PE in many countries. There is an imposing barrier to reducing the CO_2 emissions of industries and residences up to half, but it is not impossible if the proper economic growth is maintained.

(3) In order to make these targets, such as reducing CO_2 emissions and maintaining economic growth, the other important issue is how to penetrate these technologies in a vast country with 1.4B people. One method is the effective use of market forces to reduce the energy consumption, and another is an institution for promoting energy conservation and reduction of CO_2 emissions. Because energy facility concerns include long service periods (say 20–40 years), it is important to introduce better efficiency at large facilities at an early stage. From this point of view, it is also pointed out that economic planning and self control by Confucian ethic will play an important role in achieving these two targets not only by the market mechanism.

(4) The economic structure from heavy industries to light industries was forced to change due to the oil crisis in Japan. However, owing to this impact, the economic structure was changed and shifted to a more stable economic growth stage from a high economic growth stage.

At present, China encounters a crisis of climate issues under a high economic growth stage. This issue is most dire for the survival of human beings. It is expected that China will put forth its best efforts to overcome this issue and to find a solution for reconciling economic growth and climate issues as a responsible leading country in the world.

REFERENCES

Goldman Sachs, (2003). *Dreaming with BRICs:* The Path to 2050

International Energy Agency (2005). World Energy Outlook 2006

Institute of Energy and Economic, EDMC Handbook of Energy and Economic Statistics in Japan 2006

Intergovernmental Panel on Climate Change (IPCC), (2000). *Special Report on Emission Scenarios.* Cambridge University Press.

Intergovernmental Panel on Climate Change(IPCC), (2007). Mitigation, Contribution of Working Group to the fourth Assessment Report of the Intergovernmental, *Panel on Climate Change.* Cambridge University Press.

Sekimoto, H., (2005). Light a CANDLE; an innovative burn up strategy of nuclear reactors.

Tashimo, M., et al., (2008). Role of nuclear energy in environment, economy, and energy issues of the 21st century – Growing energy demand in Asia and role of nuclear –. Elsevier Ltd.

United Nations data. (2008). GDP at market prices.

United Nations data. (2008). Population

In: Energy Costs, International Developments…
Ed: Leszek Kowalczyk and Jakub Piotrowski ISBN: 978-1-60741-700-2

Chapter 9

NANO-CHANNELS FOR SOLAR ENERGY CONVERSION: ORIENTED ONE-DIMENSIONAL NANO-STRUCTURE FACILITATING CHARGE TRANSFER IN DYE-SENSITIZED SOLAR CELLS

Hong Lin**, *Yizhu Liu, Xin Li and Jianbao Li
The State Key Laboratory of New Ceramics and Fine Processing, Department of Materials Science and Engineering, Tsinghua University, Beijing 100084, P. R. China

Dye-sensitized solar cells (DSCs) considered as a promising alternative to conventional silicon-based photovoltaics have been intensively studied for the recent two decades[1,2]. Although satisfactory solar energy conversion efficiencies based on such nanocrystalline mesoscopic devices have been achieved in laboratory[3], large-scale application is still in its infancy. DSCs mostly employ liquid electrolytes which facilitate charge transfer and ensure high energy conversion efficiency. However, evaporation and leakage of the organic solvents in liquid electrolytes drastically decrease the long-term stability and make out-door operation difficult to endure. Quasi-solid and solid-state DSCs have been designed and investigated to solve this problem[4,5,6]. Nonetheless, there exists a significant disparity in photovoltaic performances of these devices compared to liquid electrolyte-based cells, mainly attributed to less facile charge transfer in quasi-solid and solid-state electrolytes.

In this short communication, a concept of oriented one-dimensional nano-channel structure enlightened by our previous results was proposed to facilitate charge transfer in DSCs. The application of such a concept in further promoting energy conversion efficiency of quasi-solid and solid-state DSCs is also discussed.

---

* Corresponding author: hong-lin@mail.tsinghua.edu.cn

1. TITANIA NANOARRAYS IN DSCS[7]

TiO_2 (anatase) nanoarrays were obtained by direct hydrothermal treatment of Ti substrates followed by ion exchange in HCl and post annealing. The as-prepared films were composed of monoclinic $H_2Ti_5O_{11}\cdot H_2O$, which has been demonstrated by SEM images, HRTEM images and SAED patterns (Figure 1).

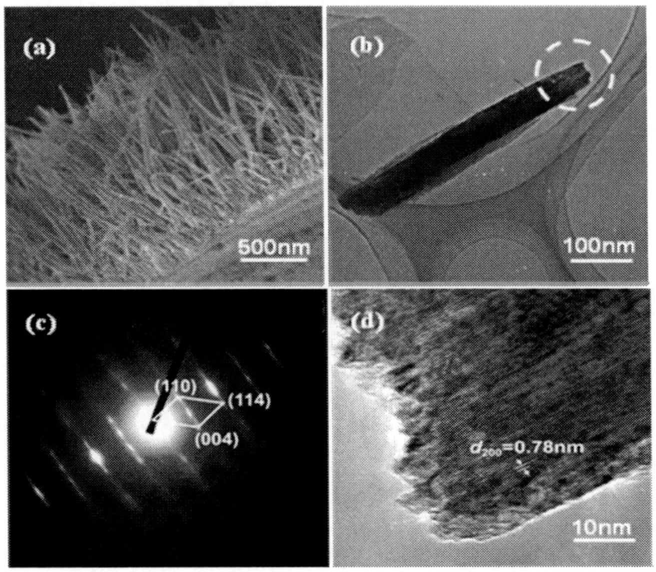

Figure 1. (a) Scanning electron microscopy image of the hydrogen titanate ($H_2Ti_5O_{11}\cdot H_2O$) nanowire array synthesized by a hydrothermal reaction at 180□ for 4 h; (b) transmission electron microscopy image of one nanowire ; (c) SAED pattern for the nanowires of the array; (d) high-resolution transmission electron microscopy image of the end of one nanowire.

A postulated mechanism of the nanowire growth is illustrated in the model as shown in Figure 2. At the beginning, the Ti substrate reacted with the NaOH aqueous solution and many $Na_2Ti_5O_{11}$ nanocrystallines were created and covered the plate. The nanocrystallines then transformed into a nucleus, from which the nanobelts begin to grow. The growth in the cross direction was inhibited because of the competitive growth from too many nuclei simultaneously, whereas the growth in the direction perpendicular to the substrate was free. As a result, the nanobelts began to grow vertically on the substrate through a dissolution–recrystallization process, during which the dislocations caused the defect strain in the cross direction and thus increased with the nanobelt growth. When the nanobelt was long enough, the strain was large enough to suppress the nanobelt into a multilayer structure and then the multilayer nanowire was formed. This mechanism was demonstrated in detail by SEM and HRTEM images of films prepared under different conditions[7]. The $Na_2Ti_5O_{11}$ nanowire array film was then immersed into an HCl aqueous solution to exchange Na^+ into H^+. The resulting $H_2Ti_5O_{11}\cdot H_2O$ nanowire array film was postannealed and transformed into a TiO_2 nanowire array film. XRD patterns below (Figure 3) showed that anatase began to form at 250□ (curve c) and $H_2Ti_5O_{11}$ phase almost disappeared completely at 450° (curve e).

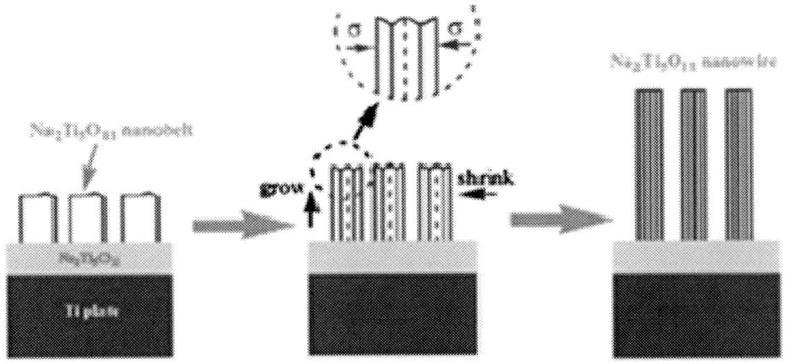

Figure 2. Mechanism of the nanowire growth

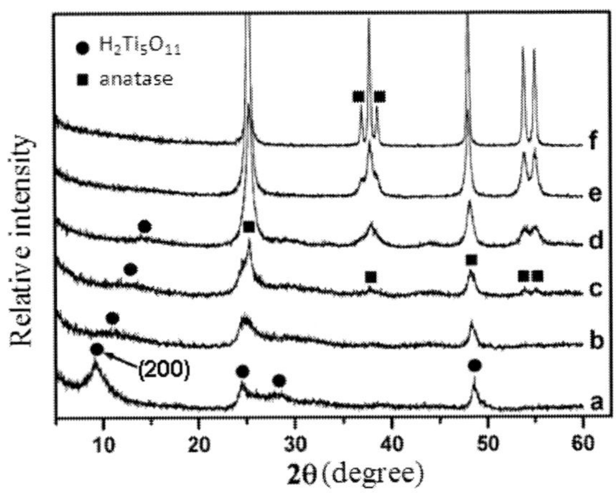

Figure 3. XRD patterns of TiO_2 nanowire array films sintered at different temperatures. (a) $H_2Ti_5O_{11}$; (b) 200°; (c) 250°; (d) 400°; (e) 450°; (f) 700°.

The nanowire thin film was shed from the plate as a membrane, pasted onto a conductive glass and assembled into a DSC. Figure 4 shows the photocurrent–voltage (I–V) characteristics of DSCs prepared using TiO_2 nanowire arrays and commercial TiO_2 powder (P25, Degauss, Germany)[8,9]. The short-circuit current density (J_{sc}) increased from 13.5 to 15.2mA/cm^2 on using nanowire arrays with a thickness of 4.1μm as the photoanode compared to the traditional nanocrystalline film (8μm thick) made of P25 TiO_2 powder despite of a much smaller thickness. The efficiency of DSC was also enhanced from 5.67% to 6.58%, which indicates that the structure of the TiO_2 thin film may affect the photocurrent of DSC.

For TiO_2 nanoparticles, there are many defects in the boundaries at the contacts between nanoparticles. These defects may act as electron traps that hamper the movement of electrons[10]. However, the oriented one-dimensional TiO_2 nanowires may provide channels for fast electron transfer while maintain the merit of high specific surface area, which allows adequate loading of sensitizer and thus exhibited even higher energy conversion efficiency with thinner films. This is potentially advantageous for quasi-solid and solid-state DSCs where the short diffusion length of injected excitons is always problematic due to poor wetting of the electrolyte.

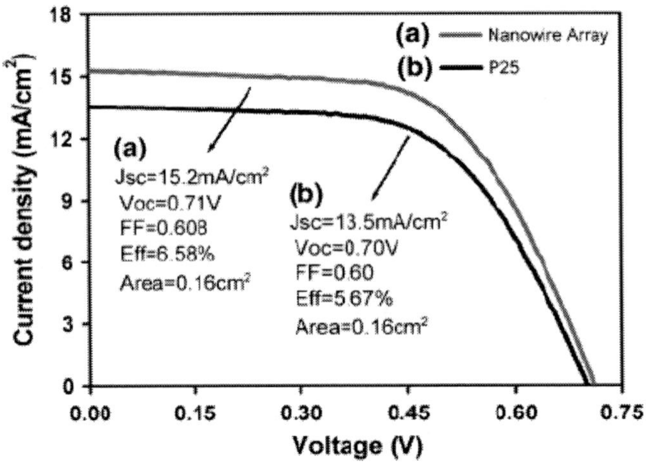

Figure 4. Photocurrent–voltage characteristics of dye-sensitized solar cells prepared using (a) a TiO_2 nanowire array synthesized by hydrothermal treatment at 180° for 4 h and postannealing at 450° for 2 h and (b) P25 TiO_2 powder.

2. DSCs with Composite Liquid Electrolytes and Ionic Liquids

2.1. Composite Liquid Electrolytes

As a feature of our effort on electrolyte modification, inorganic additives were adopted and their effects on enhancement of electrochemical properties of the electrolytes were investigated concerning with liquid electrolytes[11]. Herein, layered α-zirconium phosphate (α-ZrP)[12] was chosen and added in different weight content to iodine-based electrolyte containing 4-tert-butylpyridine (TBP) as an organic additive which has demonstrated to be able to improve markedly the open-circuit voltage and the filled factor of DSC by decreasing the black reaction[13].

Such composite electrolytes were assembled into thin-layer cells and subjected to electrochemical characterizations including electrochemical impedance spectroscopy (EIS) and cyclic voltammetry (CV). The Nyquist diagram of thin-layer cells (Figure 5(a)) showed that with the addition of α-ZrP, the R_{ct} value of thin-layer cell representing the charge transfer resistance of the counter electrode/electrolyte interface decreased markedly and reached its minimum at 9wt% α-ZrP content relative to the weight of TBP, while the R_s value representing the ohmic serial resistance (resistance of electrolyte) had almost no change. This result is supported by cyclic voltammetry characterizations (Figure 5(b)) where the diffusion coefficient of I_3^-, $D_{I_3^-}$, proportional to j_{lim}, had its maximum value also at 9wt% α-ZrP content.

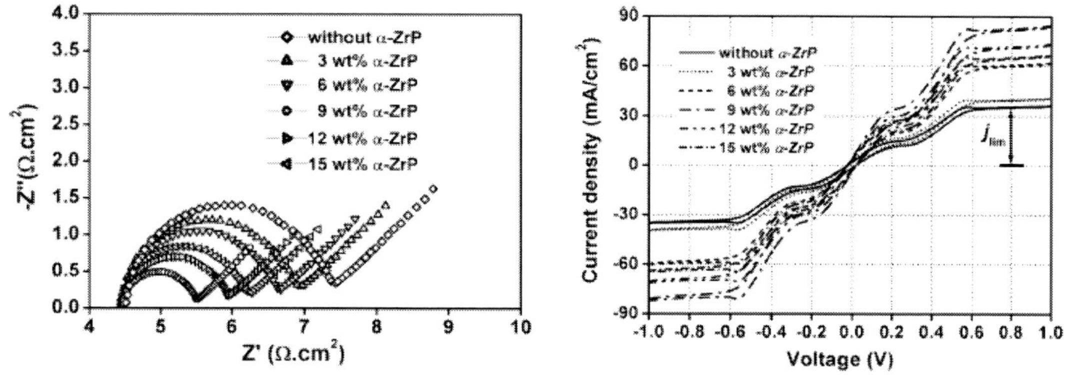

Figure 5. (a) Nyquist diagram; (b) cyclic voltammegram of thin-layer cells.

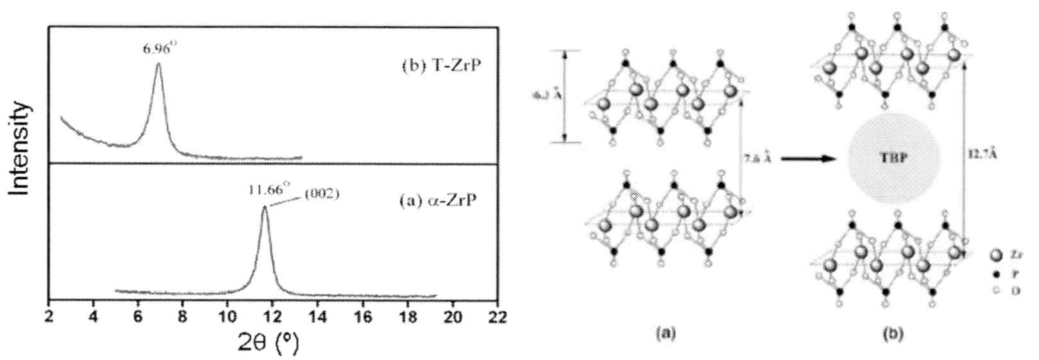

Figure 6. (a) The XRD patterns of α-ZrP doped electrolyte; (b) illustration of intercalation of TBP into α-ZrP nano-layers.

The enhancement of electrochemical properties was attributed to the intercalation of TBP molecules into α-ZrP nano-layers and was demonstrated by the XRD patterns (Figure 6(a)) of pure α-ZrP and TBP-doped ZrP (T-ZrP). The (002) diffraction peak shifted from 11.66° to 6.96°, corresponding to the change of the interlayer distance from 7.6Å to 12.7Å (Figure 6(b)). Meanwhile, the interlayer gap was enlarged into 6.4Å, exceeding both the diameter and length of I_3^- ion in acetonitrile solution[14]. Thus, I_3^- ions can transport freely in the interlayer of T-ZrP, which could pile up and form "bridge" between the two electrodes, and therefore the electrochemical properties can be greatly enhanced.

2.2 Composite Ionic Liquids and Corresponding Quasi-Solid DSCs

The significant enhancement of electrochemical properties of liquid electrolytes by layered α-ZrP is highly indicative for initiating a new field of electrolyte modification by inorganic nano-layered materials, within which the efficient modification of ionic liquids that have long been suffering low ion diffusion coefficient is expected. Thus, in our study, α-ZrP was also applied to prepare composite ionic liquids and quasi-solid DSCs were assembled thereafter based on their encouraging electrochemical properties improvements[15].

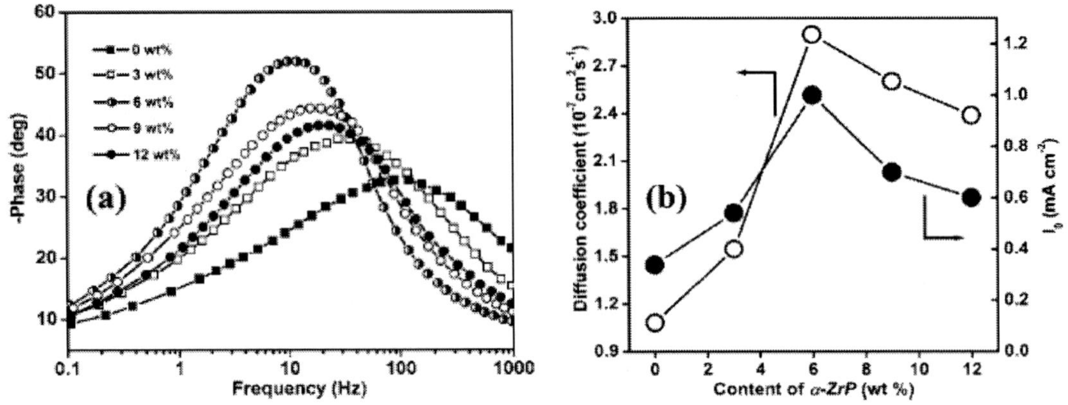

Figure 7. (a) Bode phase plots of DSCs; (b) dependence of exchange current density I_0 and I_3^- diffusion coefficient $D_{I_3^-}$ on weight content of α-ZrP.

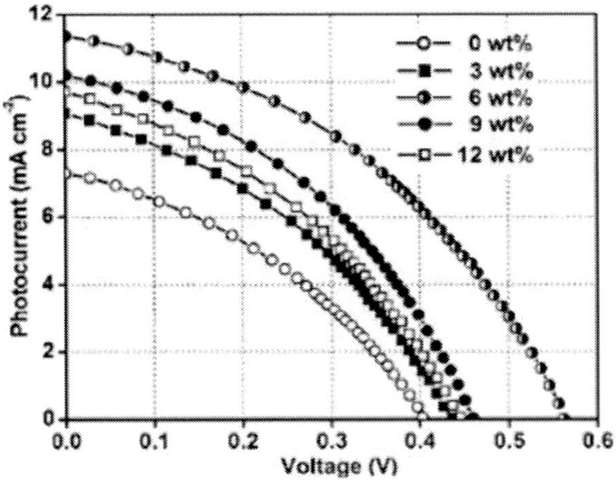

Figure 8. Photovoltaic performances of quasi-solid DSCs

α-ZrP was first added to TBP-containing 1-methyl-3-propylimidazolium dihydrophosphate (MPIDP)[16] ionic liquid at different weight contents relative to TBP, and either thin-layer cells or quasi-solid DSCs were fabricated and subjected to EIS and photovoltaic performance characterizations. The Bode phase plots of DSCs (Figure 7(a)) reflected the electron lifetime (τ_e) in TiO_2 films by the equation $\tau_e = 1/\omega_{mid} = 1/2\pi f_{mid}$, where ω_{mid} represented the characteristic angle frequency of the midfrequency (f_{mid}) peak. f_{mid} of the DSC with 6% α-ZrP decreased to its minimum of 11.13Hz by a factor of 9 compared to the DSC without α-ZrP, suggesting that τ_e is lengthened significantly. Meanwhile, the optimum of electrochemical properties of composite ionic liquids at 6wt% of α-ZrP was demonstrated by thin-layer cell characterization, when exchange current density I_0 and I_3^- diffusion coefficient $D_{I\bar{3}}$ simultaneously reached their maximum respectively [Figure 7(b)]. Photovoltaic performances of quasi-solid DSCs (Figure 8) further supported such results, that is, with the increase of α-ZrP content to 6wt %, J_{sc} and V_{oc} were both enhanced, reaching their maximum of 0.562 V and 11.38mA/cm^2, respectively. An efficiency η of 2.61% was achieved, increasing by a factor of more than 2 compared to a DSC without α-ZrP.

The effect of α-ZrP on composite ionic liquids can be interpreted in a similar way with that in α-ZrP doped liquid electrolytes. However, the optimal content of α-ZrP here is 6wt%, rather than 9wt% in composite liquid electrolytes, which is mainly due to the difference in the dispersion of α-ZrP in acetonitrile and ionic liquids.

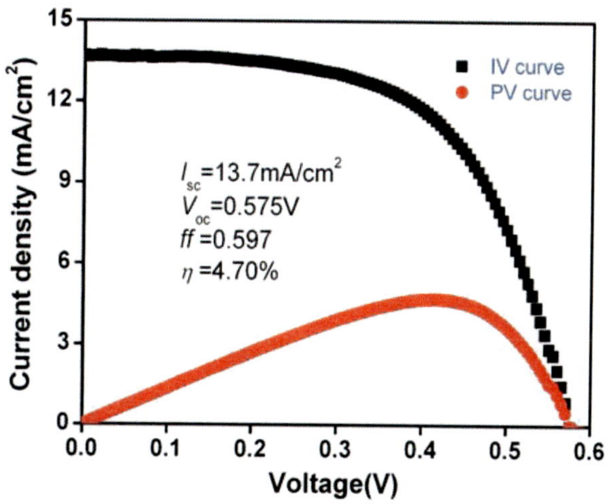

Figure 9. Photovoltaic performances of DSC with α-ZrP doped MPII

Besides MPIDP ionic liquid, several other ionic liquids were also doped with α-ZrP to prepare various composite ionic liquids, among which α-ZrP-doped 1-methyl-3-propylimidazolium iodide (MPII) exhibited 4.7% light-to-energy conversion efficiency (Figure 9).

3. Oriented One-Dimensional Nano-Channel Structure

It has been demonstrated from the examples above that oriented nano-structure does have some effect in providing oriented paths for charge transfer. Such oriented paths confine flowing charges in specific directions and significantly decrease the diffusion distance in either photoanodes or electrolytes. Charges can be more expeditiously transferred and recombination is accordingly less probable. Application of such nano-structures in quasi-solid and solid-state DSCs is especially intriguing, since drastic recombination in quasi-solid and solid-state electrolytes has been widely considered as the key point impairing their photovoltaic performances[5]. Till now, very little relative work has been carried out, and it is believed that attempts in this aspect might bring some innovative improvement for quasi-solid and solid-state electrolytes and DSCs.

References

[1] Grätzel, M. *Philosophical Transactions of the Royal Society* A, 2007, 365, 993-1005.
[2] O'Regan, B; Grätzel, M. *Nature*, 1991, 353, 737-740.
[3] Nazeeruddin, M; De Angelis, F; Grätzel, M., et al. *Journal of the American Chemical Society*, 2005, 127, 16835-16847.
[4] Murakoshi, K; Yanagida S., et al. *Chemistry Letters*, 1997, 5, 471-472.
[5] Bach, U; Grätzel, M., et al. *Nature*, 1998, 395, 583-585.
[6] Kim, S; Ko, J., et al. *Chemical Communications*, 2008, 4951-4953.
[7] Wang, W; Lin, H., et al. *Journal of the American Ceramic Society*, 2008, 91, 628-631.
[8] Hagfeldt, A; Grätzel, M. *Chemical Reviews,* 1995, 95, 49-68
[9] Wang, N., et al. *Applied Physics Letters*, 2006, 89, 194104.
[10] Adachi, M. et al. *Journal of the Electrochemical Society*, 2003, 150, G488-493.
[11] Wang, N; Lin, H., et al. *Electrochemistry Communications*, 2006, 8, 946-950.
[12] Clearfield, A. *Prog. Inorg. Chem. Res.*, 1998, 47, 371.
[13] Haque, SA; Grätzel, M., et al. *Journal of Physical Chemistry B*, 2000, 104, 538.
[14] Sakane, H., et al. *Journal of Synchrotron Radiation*, 2001, 8, 674.
[15] Wang, N; Lin, H., et al. *Applied Physics Letters*, 2006, 89, 194104.
[16] Wang, Y., et al. Chemical Communications, 2004, 17, 1938.

INDEX

A

B

C

F

G

H

I

J

K

L

M

N

O

P

Q

R

S

T

U

V

W

X

Y

Z